SO-ALA-951

C R C

PERSISTENT PESTICIDES

IN THE ENVIRONMENT

author:

CLIVE A. EDWARDS

Rothamsted Experimental Station
Harpenden, Herts, England

published by:

A DIVISION OF
THE **CHEMICAL RUBBER** CO.
18901 Cranwood Parkway • Cleveland, Ohio 44128

This book represents information obtained from authentic and highly regarded sources. Reprinted material is quoted with permission, and sources are indicated. A wide variety of references is listed. Every reasonable effort has been made to give reliable data and information, but the author and the publisher cannot assume responsibility for the validity of all materials or for the consequences of their use.

CRC MONOSCIENCE SERIES

The primary objective of the CRC Mono-science Series is to provide reference works, each of which represents an authoritative and comprehensive summary of the "state-of-the-art" of a single well-defined scientific subject.

Among the criteria utilized for the selection of the subject are: (1) timeliness; (2) significant recent work within the area of the subject; and (3) recognized need of the scientific community for a critical synthesis and summary of the "state-of-the-art."

The value and authenticity of the contents are assured by utilizing the following carefully structured procedure to produce the final manuscript:

1. The topic is selected and defined by an editor and advisory board, each of whom is a recognized expert in the discipline.

2. The author, appointed by the editor, is an outstanding authority on the particular topic which is the subject of the publication.

3. The author, utilizing his expertise within the specialized field, selects for critical review the most significant papers of recent publication and provides a synthesis and summary of the "state-of-the-art."

4. The author's manuscript is critically reviewed by a referee who is acknowledged to be equal in expertise in the specialty which is the subject of the work.

5. The editor is charged with the responsibility for final review and approval of the manuscript.

In establishing this new CRC Monoscience Series, CRC has the additional objective of attacking the high cost of publishing in general, and scientific publishing in particular. By confining the contents of each book to an *in-depth treatment* of a relatively narrow and well-defined subject, the physical size of the book, itself, permits a pricing policy substantially below current levels for scientific publishing.

Although well-known as a publisher, CRC now prefers to identify its function in this area as the management and distribution of scientific information, utilizing a variety of formats and media ranging from the conventional printed page to computerized data bases. Within the scope of this framework, the CRC Monoscience Series represents a significant element in the total CRC scientific information service.

B. J. Starkoff, President
THE CHEMICAL RUBBER CO.

This book originally appeared as part of an article in *CRC Critical Reviews in Environmental Control,* a quarterly journal published by The Chemical Rubber Co. We would like to acknowledge the editorial assistance received by the Journal's co-editors, Professor Richard G. Bond and Dr. Conrad P. Straub, both at the University of Minnesota. Professor Russell S. Adams served as referee for this article.

INTRODUCTION

The persistent pesticides, especially the organochlorine insecticides, have conferred tremendous benefits on mankind by controlling the vectors of serious human disease which have claimed millions of lives and by greatly increasing yields of many crops. As the world's population increases, so the need for food will increase, and it seems likely that our need for pesticides will increase. In recent years there have been many reports of residues of persistent pesticides in air, rainwater, dust, rivers, the sea, and in the bodies of aquatic and terrestrial invertebrates, fish, birds, mammals and human beings. The largest residues seem to occur in the tissues of animals near the top of food chains, particularly predators and carnivores, and most important, man himself. Typical values found for DDT in the various media, terrestrial and aquatic forms, and man are summarized below.

The importance of these persistent residues in the environment has not really been assessed. Although some mortality among wild animals has been confirmed, pesticides do not seem at present to threaten any species with extinction or even seriously diminish its numbers. During the 25 years that man has been exposed to these chemicals, there is little evidence of resultant illness. Nevertheless, the spread of these persistent chemicals into all parts of the environment must be a cause for anxiety, until we know much more about their possible long-term ecological effects.[298][322]

The amounts of persistent pesticides produced are colossal; in the United States alone, more than a million tons of DDT and 600,000 tons of aldrin and dieldrin have been manufactured. These chemicals disappear very slowly and much must still remain bound up in soils, mud, the atmosphere and the biota. Although the more advanced countries are rapidly becoming conscious of the potential hazards of persistent pesticides and are using smaller amounts more carefully applied, the

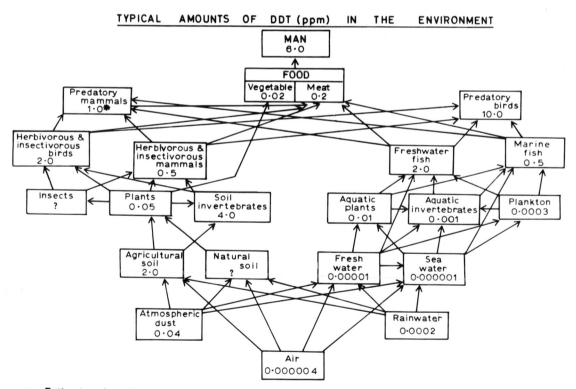

TYPICAL AMOUNTS OF DDT (ppm) IN THE ENVIRONMENT

* Estimate based on very few samples
? Insufficient data available

total world usage is increasing. In 1967, the United States produced 103 million pounds of DDT and exported 80% of this, much going to underdeveloped countries.

We still need more information about the occurrence of persistent pesticide residues in the various parts of the environment. Although considerable data exist on the levels of residues of these chemicals in man, birds and fish in North America and Europe, we have little knowledge of the amounts that occur in soil, water or the biota in Africa, Asia or South America.

This review attempts to bring together comparative data on the amounts of residues currently in the environment. It cannot be complete, but it shows where the largest residues are and how they are concentrated from the physical environment into plants and animals, and from lower organisms into the higher trophic levels of food chains. The data quoted differ greatly in accuracy and validity, sometimes due to different analytical techniques, but more often because some workers base their conclusions on a suitable number and size of samples, while others make sweeping generalizations after a very inadequate sampling program. Nevertheless, sufficient data are reviewed, from an unbiased viewpoint, to enable the reader to make an assessment of the current status of persistent pesticides in the environment, and it is hoped that this review will stimulate interest and demonstrate the need for the further research and legislation which are so urgently needed.

C.A. Edwards, June 1970
Rothamsted Experimental Station

THE AUTHOR

Clive A. Edwards is Principal Scientific Officer at Rothamsted Experimental Station, Harpenden, Herts, England. His major field of work is soil insecticides and ecology of soil invertebrates.

He has earned an M.Sc. degree from Bristol University and M.S. and Ph.D. degrees from University of Wisconsin. Dr. Edwards has written over 70 scientific papers and also received the following fellowships: Kellogg Foundation Fellow, National Science Foundation Fellow, and Senior Foreign Scientist N.S.F. Fellow.

TABLE OF CONTENTS

Persistent pesticides are not new; as long ago as 70 A.D., Plinius recommended arsenic as an insecticide and the Chinese regularly used arsenic sulfide for this purpose in the late sixteenth century.[207] With the expansion of agriculture and medicine during the last hundred years the need for effective insecticides to control pests of crops, animals and human beings increased. The earlier insecticides included various inorganic compounds which contained lead, antimony, arsenic, mercury, selenium, sulfur, thallium, zinc, and fluorine as the active ingredients. These compounds, although not very toxic to insects, were very persistent; so sprayed crops sometimes retained sufficient arsenical residues to be potentially harmful to the consumers, and crops were sometimes damaged by residues which accumulated in soils.[18] Fungicides containing sulfur, copper, and mercury have been another source of persistent chemical residues in soils.

These chemicals had limited usage, however, and the persistent pesticides were not widely used until the discovery of DDT and other chlorinated hydrocarbon insecticides during and after the second world war. The chlorinated hydrocarbons kill insects and many other arthropods very readily, so small doses are adequate, and as these insecticides are not very toxic to vertebrates it seemed unlikely that they would harm animals which are not pests. A single dose of insecticide applied to an arable soil often controlled pests for several seasons and it seemed that most soil pest problems had been conquered. Although these insecticides were useful in agriculture they were even more valuable against pests that carried diseases of animals and human beings. Arthropods which transmit diseases include lice (which spread typhus, trench fever, relapsing fever), fleas (which spread plague), ticks (which spread Texas fever), mosquitoes (which spread malaria, yellow fever, elephantiasis), and mosquitoes and tsetse flies (which spread sleeping sickness). There is no doubt that the use of chlorinated hydrocarbon insecticides has saved millions of lives and greatly increased food production.

Chlorinated hydrocarbons were known to be very persistent, but until the early 1950's there was little anxiety as to long-term hazards caused by their use. There was evidence that large residues in soil could be phytotoxic; small quantities of some were found in plant and animal tissues and in cow's milk, and sometimes fish were killed when water was sprayed in anti-malarial campaigns, but these were accepted as slight but unavoidable hazards and of little concern.

During the 1950's and in the early 1960's there were reports of large residues of these insecticides in soils[41 96 102 165] and small amounts in water and at the bottom of streams.[228] Dead birds were sometimes found close to sprayed fields and woodlands, or fields planted with insecticide-treated seed,[216 257] and dead fish were seen on the surface of water after spraying operations.[213 295] There were indications that chlorinated hydrocarbon insecticides were not only stored in invertebrate and vertebrate tissues but were also concentrated into the upper trophic levels of food chains.[10 135] These discoveries began to cause concern about possible long-term effects of the large quantities of insecticides being used, sometimes indiscriminately, and this culminated with the publication of *Silent Spring*.[37] This book, by greatly exaggerating potential hazards of persistent pesticides, focused much more attention on the problem, no doubt helping towards awareness of the need for research, and thus indirectly contributing to our knowledge of the present status of pesticide residues in the various sections of the environment.

The Fish and Wildlife Service, the U.S. Department of Health, Education, and Welfare, and the U.S. Department of Agriculture in the United States, and the Nature Conservancy, and the Ministry of Agriculture, Fisheries and Food in Great Britain, and equivalent organizations in many other countries are now monitoring the residues of persistent pesticides in the physical and biological elements of the environment, and the effects of these residues on the biota. Many chemists and ecologists are also working on these problems, either independently or in association with government organizations.

Nevertheless, the monitoring of residues in various parts of the physical environment is still very inadequate, particularly in Great Britain. Major rivers in the United States are now being monitored annually for pesticide residues at a large number of sampling stations, but in Great Britain only one or two small scale surveys have been made. The monitoring of soils for pesticides is much more sporadic, both in the U.S. and Great Britain, and so far has been mainly confined to agricultural soils, so that there is very little information on the amounts of pesticides that occur in the large areas of untreated land; such information is essential for adequate assessment of the hazards caused by persistent pesticides. There is considerable information on amounts of pesticide residues in the biota, but this is scattered and uneven, so that there have been many more investigations into residues in fish, birds, and birds' eggs than in other vertebrates and invertebrates. Residues have been found in the air and rainwater but as yet there is no regular monitoring of these. A considerable amount of data on residues in the human diet, and in bodies of human beings in many parts of the world, has now been collected and it is now much more possible to assess possible hazards to man. This paper attempts to summarize the results of some of the more important residue surveys, and experimental studies of residues in both the physical and biological sections of the environment, in order to assess the present state of knowledge of the status of pesticide residues in the environment. It is beyond the scope of this paper to do more than point out the possible hazards of the present amounts of pesticides; the biota may well adapt to such hazards if they continue.

During the past five years there have been some very relevant reviews and symposia on the persistence of pesticides in the environment; the first of these, *Pesticides and the Living Landscape*,[229] was a much more balanced review of the influences of pesticides in nature than Carson's[37] biased and controversial book. In 1964, a comprehensive annotated bibliography entitled *Pesticides in Soils and Water* was published by the U.S. Department of Health, Education, and Welfare;[261] in the following year an excellent review was published on *Residues of Chlorinated Hydrocarbon Insecticides in Biological Material*, with particular reference to residues in plant material,[184] and Moore[191] reviewed the effects of pesticides on birds in Great Britain. These reviews were followed in 1965 by a symposium entitled *Re-*

search in Pesticides[39] which contained several papers on persistence of pesticides in the environment. In 1966 there were three symposia: the first, organized by the Soil Society of America, was published under the title *Pesticides and Their Effects on Soils and Water,*[6] and the proceedings of another symposium, arranged by the American Society for the Advancement of Science, had the title *Agriculture and the Quality of Our Environment;*[19] both of these included several informative papers as did a volume, *Organic Pesticides in the Environment* which contained papers read at an American Chemical Society meeting in New Jersey 1965.[105] Papers read at an international symposium organized by the British Ecological Society were published with the title *Pesticides in the Environment and Their Effects on Wildlife*[192] and in the same year a book, *That We May Live,*[285] written by a long-term chairman of the U.S. House of Representatives' Appropriations Subcommittee for Agriculture, was really a refutation of many of Carson's[37] implications concerning pesticides. Edwards[76] reviewed the state of information concerning soil insecticide residues, particularly emphasizing factors influencing their persistence in soil.

Moore[194] published a paper, *A Synopsis of the Pesticide Problem*, which dealt with ecological aspects of pesticide usage, and a review by Stickel[237] on *Organochlorine Pesticides in the Environment* also briefly discussed the ecological aspects of insecticide persistence. Newsom[203] reviewed the consequences of insecticide usage on non-target organisms, but his review, although good, was limited in scope and coverage. A popular book by Mellanby,[186] *Pesticides and Pollution*, discussed some of the more elementary aspects of pesticide pollution. The Department of Radiation Biology and Biophysics at the University of Rochester organized a symposium under the title of *Chemical Fallout—Current Research on Persistent Pesticides*[188] which contained some excellent papers, especially one, *Organochlorine Insecticides and Bird Populations in Britain.*[220] Johnson[147] published a good review on *Pesticides and Fishes*, with an excellent bibliography, and Robinson[221] summarized the uptake of chlorinated hydrocarbon insecticides into human beings, emphasized the deficiencies in sampling and analytical techniques, and produced a mathemat-

ical model to account for the metabolism of pesticides in the human body.

Thus there is no lack of specialized reviews or proceedings of more general symposia concerning pesticides in the environment. However, there are no reviews which cover the entire subject other than the books by Carson[37] and Rudd,[229] and these made no attempt to assess the amounts of residues in the environment at the time that they were written, possibly because available information was sparse. The symposia, although contributing useful individual articles, did not cover the whole field, or summarize fully the routes by which residues were carried between the different components of the environment. For this reason, this review attempts to present recent data on the amounts of residues in all compartments of the environment, and although the data are still inadequate, to show how they are concentrated from the physical to the biological parts of the environment and into the upper trophic levels of food chains. It is hoped that this review will demonstrate the inadequacies of our knowledge of residues in some media and organisms, and particularly how little is known about the amounts of residues which exist in whole continents, where chlorinated hydrocarbon insecticides have been used for many years.

This review will quote much data in an attempt to establish the current amounts of residues in the environment, but it must be emphasized that the data quoted have been obtained by a large number of investigators, often using different analytical techniques, sampling procedures, and care in doing their analyses. Frequently the data are given in a form which is very difficult to summarize or compare with that of other workers. A very important point is that as analytical methods have improved and become more sensitive, the possible misinterpretations that can be placed upon the data they provide have multiplied. This is particularly true of the gas-liquid chromatographic techniques which rely on relating the position of a peak traced on a chart with one given by a particular compound which is to be estimated. Many different compounds can produce peaks very close together; when peaks are large, well-defined and clearly separated, they usually provide a valid estimate of the amount and kind of residue present, particularly if they can be con-

FIGURE 1

PESTICIDE CYCLING IN THE ENVIRONMENT

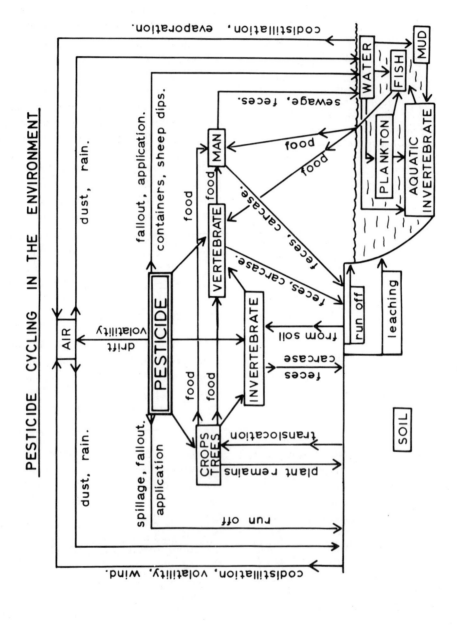

firmed by paper chromatography or other analytical techniques. In analysis of extremely small amounts of residues such as those in air or rainwater, there is much more possibility of error, not only by misinterpreting peaks but also from contamination of the analytical equipment with small traces of pesticides or related compounds. Chlorinated hydrocarbon insecticides have often been reported from materials which could not possibly contain them, such as in soil samples stored in sealed containers since before the second world war, or in food that was canned more than fifty years ago; thus wherever possible, more than one analytical method should be used to confirm the presence of the suspected residue. For the purpose of this review, all published data have been accepted as valid, although some may well be suspect, and this should be remembered when the data are quoted or discussed. What is more, explanations of contaminations of soil, water, or the biota with pesticides may not always be complicated or ecological. It has been suggested that the minute amounts of chlorinated hydrocarbon insecticides reported from fish and vertebrates in the Antarctic may be accounted for by insecticides imported into the area with various expeditions. Such possiblities should be borne in mind when interpreting residue data.

The other point that must be made is that residues are usually analyzed at one point in time, and such data can easily be misleading. For example, many workers take a single set of samples from the physical or biological environment and determine the residues they contain. Such a spot check may have very little value, however, and may be little related to the average amounts that occur in that medium. In the physical environment these residues are probably the remains of a larger dose, whereas in the biota they can be the result of a long process of absorption and concentration. The whole system of cycling of pesticides in the environment is essentially dynamic and, for a meaningful interpretation, must be seen and studied as such. The most difficult thing to assess from data currently available is not what the present situation is, but what changes will occur if certain actions, e.g., a ban on a particular pesticide, are taken. What we need to know is how rapidly residues will disappear from the different compartments of the environment when they are no longer renewed. We also need to know if the residues that exist are of any real importance or if particular organisms living in the environment can successfully adapt to such chemicals. Our environment is already full of alien chemicals; is the addition of these pesticide residues important or not?

An attempt to summarize the cycling of pesticides in the environment is made in Figure 1 (see page 10), which illustrates the movements of residues through the various compartments of the environment and which should be referred to when the amounts of residues are discussed in the different sections.

OCCURRENCE AND PERSISTENCE OF PESTICIDES IN THE PHYSICAL ENVIRONMENT

Pesticides in Soils

Whether pesticides are applied as aerial sprays, dusts to foliage, or directly to soil, there is little doubt that large amounts of them ultimately reach the soil which acts as a reservoir for these persistent chemicals until they move into the bodies of invertebrates, pass into air or water, or are broken down. The amounts of residues that have been found in soil in surveys are summarized in Tables 1 and 2 (see pages 12, 13), but with the exception of Lahser and Applegate,[161] they all refer to the agricultural soils which have been treated with pesticides. There is a pressing need for information on the amounts of insecticide residues which exist in woodland, grasslands, and other natural areas that have never been treated with any insecticides.

Woodwell and Martin[291] reported that residues of DDT in a forest stand in New Brunswick increased from 0.169 kg/ha in 1958 to 0.628 kg/ha in 1961, and Yule[296] found somewhat smaller amounts of DDT in another forest stand in New Brunswick in 1967 and 1968.

TABLE 1

RESIDUES OF DDT IN SOIL SURVEYS

Location	Reference	No. of sites	Cropping	Residues DDT & related compounds (pp 10^{-6})*	
				Max.	Mean
Great Britain	Wheatley et al. 1962 (284)	21	Potato fields	0.96	0.20
Great Britain	Edwards 1969 (79)	5	Orchards	11.4	7.1
Great Britain	Davis 1968 (55)	10	Arable	0.8	0.3
		2	Orchards	17.2	9.7
Canada	Harris et al. 1966 (116)	4	Sugar beet	1.5	0.4
		5	Pasture	2.6	0.5
		6	Corn	3.7	1.2
		9	Cereals	5.1	1.4
		4	Greenhouse	2.6	1.5
		5	Tobacco	5.1	3.2
		11	Vegetables	47.6	9.5
		8	Orchards	131.1	61.8
Canada	Saha et al. 1968 (230)	20	Legume & cereal	0	0
Canada	Duffy & Wong 1967 (68)	24	Roots	17.1	1.7
		11	Vegetables	6.7	2.0
		3	Strawberries	1.81	1.0
		6	Orchards	3.6	2.4
U.S.A. (New Jersey)	Ginsburg & Reed 1954 (102)	11	Orchards	73.0	42.8
		8	Potato fields	5.1	2.3
		10	Corn	6.5	4.0
U.S.A. (Missouri, Indiana, Ohio, Michigan, Wisconsin, Iowa, Illinois, N. Dakota.)	Lichtenstein 1958 (165)	14	Orchards	106.0	29.4
		24	Arable fields	4.6	1.2
U.S.A. (Michigan)	Fahey et al. 1965 (89)	227	Turf & cultivated	87.3	2.9
U.S.A. (Texas)	Lahser & Applegate 1966 (161)	3	Cotton	2.6	2.4
		5	Desert & prairie	2.3	1.6
U.S.A. (Arizona)	Ware et al. 1968 (274)	12	Alfalfa	4.9	1.4
U.S.A. (Eastern states)	Seal et al. 1967 (233)	5	Peanuts	0.7	0.3
		25	Potatoes	7.0	2.8
		19	Carrots	12.8	3.7
U.S.A. (Western states)	Trautman 1968 (253)	41	Arable	4.6	0.7
U.S.A. (Southern & Midwestern states)	U.S.D.A. 1968 (260)	27	Soybeans	7.5	1.7
				pounds per acre †	
U.S.A. (Indiana, New Jersey, Washington)	Chisholm et al. 1951 (40)	31	Orchards	116.0	37.1
U.S.A. (Indiana)	Murphy et al. 1964 (200)	35	Orchards	362.0	148.0

* mg/kg

† 2 lb per acre in a six inch soil depth approximates to 1 (pp 10^{-6})

TABLE 2

RESIDUES OF CYCLODIENE INSECTICIDES & RELATED MATERIALS IN SOIL SURVEYS.

RESIDUES (pp 10^{-6})*

Location	Reference	No. of sites	Cropping	γ BHC Max.	γ BHC Mean	Aldrin Max.	Aldrin Mean	Dieldrin Max.	Dieldrin Mean	Chlordane Max.	Chlordane Mean	Heptachlor & heptachlor epoxide Max.	Heptachlor & heptachlor epoxide Mean
Great Britain	Wheatley et al. 1962 (284)	21	Potatoes	-	-	0.12	0.02	0.41	0.09	-	-	-	-
Great Britain	Edwards 1969 (78)	12	Carrots	-	-	-	-	1.47	0.67	-	-	-	-
		5	Cereal	-	-	-	-	0.40	0.02	-	-	-	-
Great Britain	Davis 1968 (55)	10	Arable	0.01	t	0.7	0.15	0.7	0.15	-	-	-	-
Canada	Duffy & Wong 1967 (68)	24	Roots	-	-	2.13	0.49	4.04	0.47	0.48	0.03	0.73	0.04
		11	Vegetables	-	-	2.50	0.43	1.14	0.25	0.86	0.12	1.39	0.16
		3	Strawberries	-	-	0.01	0.01	0.02	0.01	-	-	-	t
		6	Orchard	-	-	0.15	0.03	0.16	0.04	-	-	0.02	-
Canada	Harris et al. 1966 (116)	4	Sugar beet	-	-	-	-	-	-	-	-	-	-
		5	Pasture	-	-	0.2	0.04	1.1	0.2	-	-	-	-
		6	Corn	-	-	0.5	0.02	0.9	0.3	0.1	0.02	-	-
		9	Cereals	-	-	0.5	0.06	1.1	0.1	-	-	-	-
		4	Greenhouse	-	-	-	-	0.4	0.1	-	-	-	-
		5	Tobacco	-	-	0.2	0.04	0.5	0.3	0.2	0.06	0.2	0.06
		11	Vegetables	-	-	2.1	0.33	1.6	0.8	0.6	0.08	0.2	0.02
		8	Orchards	-	-	-	-	-	-	-	-	-	-
Canada	Saha et al. 1968 (230)	20	Legume & cereal	-	-	0.05	0.01	0.3	0.05	0.02	t	0.05	t
U.S.A. (Michigan)	Fahey et al. 1965 (89)	227	Turf & cultivated	1.4	0.04	-	-	2.2	0.03	120.0	1.2	1.6	0.03
U.S.A. (Illinois)	Decker et al. 1965 (59)	35	Corn	-	-	-	-	1.22	0.50	-	-	-	-
U.S.A. (Texas)	Lahser & Applegate 1966 (161)	3	Cotton	0.6	0.26	-	-	-	-	-	-	-	-
		5	Desert & prairie	0.3	0.2	-	-	-	-	-	-	-	-
U.S.A. (Eastern states)	Seal et al. 1967 (233)	5	Peanuts	-	-	-	-	0.2	0.15	-	-	0.1	0.08
		25	Potatoes	-	-	-	-	0.2	0.1	-	-	0.26	0.16
		19	Carrots	-	-	-	-	0.26	0.2	-	-	0.01	t
U.S.A. (Western states)	Trautman et al. 1968 (253)	41	Arable	t	t	0.15	t	1.52	0.06	-	-	-	-
U.S.A. (Southern & midwestern states)	U.S.D.A. 1968† (260)	27	Soybeans	-	-	0.18	0.02	0.31	0.08	1.11	0.24	0.16	0.02

t = trace of residue

* mg/kg

† also contained endrin 1.17 (pp 10^{-6}) max 0.1 (pp 10^{-6}) mean.

These seem to be the only data available on residues in woodland soils, and even these are from stands that had been regularly sprayed.

The most common residues in agricultural soils are of DDT and related compounds which reach a mean, for all the surveys, of several parts per million; few of the soils surveyed contained no detectable DDT residues. The soils which contained the largest amounts of DDT residues were the orchard soils, although some fields that had principally grown vegetables also contained large amounts of residues. The next most common residue to DDT was dieldrin, but this insecticide was found in much smaller quantities than DDT, although possibly more frequently. This is surprising because Edwards[76] made calculations, from all the published data available at the time, which showed that although DDT was the most persistent insecticide in soil, it was only slightly more persistent than dieldrin. During 1962-64, before the use of aldrin and dieldrin was restricted in Great Britain, it was estimated by Strickland[245] that 262 tons of DDT (and "Rhothane") were applied to 262,000 acres of arable land, whereas 163 tons of aldrin and dieldrin (aldrin breaks down to dieldrin) were used on 715,000 acres. In the United States, the annual statements of the U.S. Department of Agriculture show that 23,660 tons of DDT were used in 1964-65, compared with 36,000 tons of the aldrin and dieldrin group of insecticides. The most likely explanations for the larger residues of DDT in soil seem to be the longer period of usage of DDT compared to that of aldrin or dieldrin, and its use for controlling mosquitoes and forest pests by aerial treatments. At present, DDT seems to present a greater potential hazard in the environment than any of the other persistent pesticides because of its greater persistence and affinity for fatty tissues.

In Great Britain, very few other residues have been found in soil, and even these have been little more than traces; this is certainly because of the very small quantities of other kinds of chlorinated hydrocarbons that have been used there. Small quantities of chlordane have been used for killing earthworms; heptachlor has been used as seed dressings, and endrin as a spray to kill black currant gall mite, but the total quantities of all these insecticides used have been infinitesimal compared with those of

DDT and dieldrin. Table 2 (see page 13) shows clearly that quite large residues of chlordane and heptachlor are found in agricultural soils in the United States, and probably when soil surveys are extended to the southern United States, residues of endrin will also be found to be common, because large amounts of endrin are used on cotton crops. Significant endrin residues were found in soils of the Mississippi River Delta (U.S.D.A. 1966).[259]

There is little doubt that the large residues of DDT and dieldrin in soils have been partly due to misuse of these insecticides. Aerial spraying and annual 'insurance' treatment of large areas, irrespective of whether or not pests were present, have contributed greatly to these residues, and the practice of combining persistent insecticides with fertilizers has often caused use of insecticides when they were not really needed.

The persistence of chlorinated hydrocarbon insecticides in soil is very variable, particularly in relation to the type of soil to which the insecticide is applied. It is possible, however, to calculate an average halflife for any insecticide, and Edwards,[76] who calculated such halflives for persistent insecticides from published data, found no halflives greater than four years. The breakdown of chlorinated hydrocarbon insecticides in soil is not, however, exponential, so the use of the term halflife is not really valid. Nevertheless, the breakdown curve may superficially appear to be exponential[150] although it is made up of several distinct components[76] which differ in importance with the climate, season, or soil type so that the halflife of a particular insecticide differs greatly for each situation.

If the average halflife is known approximately, it is possible to calculate the likelihood of an insecticide accumulating in soil after regular treatments.[109] For halflives of up to one year, assuming that breakdown is close to exponential, the residues occurring in soil will not be more than twice the annual addition whether the increments are added once a year or four times. If the halflife is four years, then the maximum accumulation would be about six times the annual dosage, and if it is as much as ten years, the upper limit would be no more than 15 times the annual dose. This was demonstrated by Decker,[59] who calculated the expected amounts of residues of aldrin in 35 cornfields in Illinois after annual treatments for up to ten years, then sampled the fields and analyzed the residues. Agreement between expected and actual residues was remarkably close. It seems likely that where residues accumulate, there have usually been treatments of more than 2 lb of insecticide per acre annually; this is certainly true of orchards which may have up to six annual treatments with DDT.

Experimental treatment of soils followed by regular analyses (e.g., Nash and Woolson[201]) have confirmed[76] that DDT and dieldrin persist longest in soils, followed by lindane, chlordane, heptachlor, and aldrin in order of decreasing persistence. There is little information on the persistence of endrin in soil,[231] but the work of Nash and Woolson indicates that it persists for about as long as DDT, and toxaphene persists for a similar time to dieldrin.

Edwards[76] concluded that the chemical structure of an insecticide, and its resultant intrinsic stability, is the most important single factor affecting the time it persists in soil. The more volatile an insecticide, the shorter the time it persists in soil,[112] and the effectiveness of an insecticide tends to be inversely proportional to its water solubility.[113]

The concentration of an insecticide is also important, proportionately more disappearing from a small dose than a large one, although a larger gross amount of insecticide may disappear from a large dose.[76] The formulation applied also influences persistence, granules persisting longer than emulsions which, in turn, persist longer than miscible liquids and wettable powders. The type of soil treated with an insecticide greatly influences how long residues persist;[18][286] for instance, heavy clays retain insecticides much longer than lighter sandier soils. Most insecticides are adsorbed on to soil fractions; however, they may become bound to the soil so tightly that they are non-toxic to either insects or plants[80] and cannot be taken up into the bodies of these organisms. When an insecticide is applied to the surface of mud, it gradually penetrates into the interior of the mud;[11] this must also occur with soil. The content of organic matter in soil seems to be the most important single factor influencing how long insecticide residues persist in soil, although the amount of clay minerals in soil can also be correlated with the persistence of residues. Edwards et al.[80] postulated that organic

matter in soil influenced insecticide persistence in a curvilinear manner, and this has since been confirmed by Harris[114] and Hermanson and Forbes.[124] There is evidence that organophosphorus insecticides persist longer in acid soils than in alkaline ones, and there are indications that the same is true for chlorinated hydrocarbon insecticides. The amount of mineral ions, such as Fe^{++}, Al^{+++} and Mg^{++}, present in a soil can also affect the adsorption and persistence of the insecticide.[94][98][99]

Climatic factors greatly influence the persistence of insecticides in soil; increased temperatures accelerate the loss of insecticide by increasing the rate of conversion to other compounds,[172] volatilization and desorption,[115] but coincidentally, if the soil becomes dry, the loss of insecticide is retarded. Rain does not leach the relatively insoluble chlorinated hydrocarbon insecticides very far into soil, and although Edwards[78] and Thompson et al.[252] showed that dieldrin could penetrate down through cracks in soil, they found that no more than 2% of the amount applied to soil reached drainage water. Edwards et al.[84] showed that the insecticides are much more likely to reach rivers and ponds carried on soil particles in water that runs off the surface of agricultural land. Soil moisture affects persistence of insecticides in another way by influencing the adsorption of the insecticides; in dry soils the non-ionic, but polar insecticides, which compete with the highly polar water molecules for sites in soil particles, are tightly adsorbed, whereas in wet soils they are released, are insecticidally active, and more readily broken down or physically removed.

Soil insecticides are lost much less rapidly from soils with a cover crop,[170] probably because the air movement over the soil surface is lessened, and there is less volatilization or co-distillation. Insecticide residues which fall on to the soil surface, but are not cultivated into the soil, disappear as much as ten times faster as those which are thoroughly cultivated into the soil;[170] this is fortunate, because it means that residues falling on to the soil surface accidentally, can usually be expected to disappear faster than when cultivated in.

It is becoming increasingly obvious that soil microorganisms play an important role in degeneration of insecticide residues in soil. Lichtenstein and Schulz[167] showed that sterilizing soil with heat retarded the breakdown of aldrin and heptachlor subsequently applied to it, and Matsumura and Boush[185] found that two species of *Trichoderma*, two *Bacillus* spp., and six species of *Pseudomonas* could degrade dieldrin in soil. Tu et al.[255] showed that most of 92 pure cultures of soil microorganisms had some capacity for converting aldrin to dieldrin, although some required a period of adaptation before they could degrade aldrin. Getzin and Rosefield[101] demonstrated that several organophosphorus insecticides decomposed much faster in autoclaved soil than in gamma-irradiated soil. They extracted a heat-labile water-soluble substance from several radiation-sterilized soils that accelerated the breakdown of insecticides in the soil. Guenzi and Beard,[107] demonstrated that DDT broke down to DDD much faster in soil under anaerobic conditions than under aerobic conditions, particularly when organic matter in the form of chopped alfalfa was added. They made the interesting suggestion that perhaps the breakdown of DDT in soil could be accelerated by creating anaerobic conditions in soil, as for example by flooding.

The potential hazards of insecticide residues in soils are probably not great, but the soil does constitute a reservoir of these residues, which may move into other parts of the environment. Fortunately, residues do not concentrate from soils into plant tissues (Table 15, see page 34); but, nevertheless, the small quantities in plant tissues which are used for human food may be undesirable. The residues concentrate into the bodies of invertebrates that live in soil, and they can be transported into the bodies of higher organisms from these; we do not know at present if this is important. Pesticide residues in soils greatly alter the numbers of invertebrates that live in soil; many soil animals, particularly insects and their larvae, are killed, but the numbers of some animals in the microfauna increase greatly, because their principal predators are killed.[75][77][81-83] In agricultural soils these changes in numbers of invertebrates are probably of no importance in terms of soil fertility; Edwards[75] showed there was no change in turnover of decaying plant material after treatment of arable soil with aldrin or DDT, but in forest and woodland soils pesticide residues may be a serious hazard in delaying the turnover of decaying material and returning it

to the soil, as a result of the effects of these residues on soil invertebrate populations. We urgently need to know how pesticide residues in woodlands indirectly affect the cycling of organic matter in soils. A further hazard of soil pesticide residues is that, after continued exposure to small quantities of insecticides, pests may become resistant to the insecticide and require ever increasing quantities of the same, or alternative chemicals, to control them.

It seems probable that progressively less chlorinated hydrocarbon insecticides will be used in soil in the future; in Great Britain there has been a voluntary restriction[190] placed on the use of aldrin and dieldrin, which led to an estimated fall of 25% in the tonnages of aldrin, dieldrin, and DDT used. Recently,[190a] the use of these chemicals has been even further restricted. Sweden, Denmark, and Norway have also restricted the use of chlorinated hydrocarbons, particularly DDT. In the U.S., alternative insecticides, which are less persistent than the chlorinated hydrocarbons, are being used to control many pests. Some states have banned particular chlorinated hydrocarbons, and the U.S. Dept. of HEW has recommended that DDT should be banned in two years' time. Thus it is likely that soil residues of persistent insecticides will begin to diminish, although even if no more persistent insecticides were used, there would certainly still be insecticide residues in soil for several decades to come.

Pesticides in Air and Rainwater

Detection of very small quantities of pesticides in air and rainwater requires extremely sensitive analytical techniques and was impossible before the development of gas-liquid chromatography. Even with this technique, it is essential to ensure that the pesticide is really present and that the peaks on the trace are not merely spurious ones due to compounds closely related to pesticides. Small quantities of pesticides were discovered in the air over peach-growing areas in Georgia and South Carolina in the United States[20] and in the same year Wheatley and Hardman[282] reported small amounts of chlorinated hydrocarbon insecticides in rainwater over central England. Wheatley and Hardman[282] were trying to account for pesticide residues in the soil on untreated plots of land, but the amounts they found in rainwater samples were much too small to do this. Weibel et al.[279] reported much larger amounts of DDT in rain collected in Ohio, U.S.A., than those found by Wheatley and Hardman,[282] but there were larger amounts of dieldrin in the English samples. One of the difficulties in analyzing pesticides in air or rainwater is that of separating particulate matter from the vapor or liquid phase. Cohen and Pinkerton[43] analyzed dust extracted from rainwater for insecticides and found many more insecticides in it than in the rainwater (Table 3, see page 18). Wheatley and Hardman[282] believed that the residues in their rainwater samples came from codistillization of pesticides from the surface of treated soils, but it seems likely that some of the residues in the rainwater must have come from dust which the rain picked up from the atmosphere, particularly because chlorinated hydrocarbons have an affinity for dust. Abbott et al.[1] sampled air over London, extracted the pesticides and estimated residues by gas-liquid chromatography; they claimed that they were not merely extracting dust out of the air, but that there was evidence that some of the pesticides extracted volatilized back into the air. The amounts of BHC they found in air were less than the previous workers found in rainwater, but they found comparable amounts of dieldrin and rather more DDT. Tarrant and Tatton[249] made a further survey of rainwater from seven sites scattered over the British Isles and reported larger amounts of DDT and BHC than Wheatley, but smaller amounts of dieldrin, and the authors suggested that smaller dieldrin residues might be due to the voluntary restriction on the use of dieldrin from 1964.[190] An alternative explanation was that between 1965 and 1968, smokeless fuel was used much more, so there were fewer dust particles in the atmosphere to collect the pesticides.

Tabor[248] also studied the amounts of pesticides in air, but he compared amounts in air over agricultural communities with those in air over an urban one; it should be noted in this context that as much as a third of the amounts of insecticides used in agricultural areas in the United States may be used in urban areas for controlling mosquitoes and other pests. Tabor reported variable amounts of insecticides in the air over the agricultural communities, particularly in one which had much larger amounts of DDT than any of the others, but the mean levels were rather low. By comparison, the air over an eastern city contained amounts of DDT comparable with those reported by Tarrant and Tatton[249] from Great Britain, but contained no other pesticide residues. If the sparse information available on the amounts of residues in air and rainwater (Table 3) is compared, it suggests that the main difference between the United States and Great Britain is that small residues of chlordane, heptachlor, and toxaphene have been found in samples from the U.S. but not in those from Great Britain; this corresponds with the national usage of these pesticides. Tarrant and Tatton[249] suggested that organochlorine residues may be transported long distances by winds in the atmosphere, and there was some support for this hypothesis from dust samples taken from the air in the northeast trade winds. Riseborough et al.[215] found appreciable quantities of DDT and dieldrin in dust particles from the air over Barbados (7.8×10^{-14} g/m^3) and calculated from the amounts they found that 600 kg of pesticides per year would reach an area between the equator and 30°N latitude covering 1.94×10^{17} cm^2. No pesticide residues were found in air samples collected at 12 stations between 17°N and 18°S near 180° longitude in the Pacific during the summer of 1967, but the total pesticide contents of dust in air over a pier at La Jolla, California ranged from 6 to 270×10^{-12} g/m^3, with an average value about 1,000 times greater than that of air over Barbados. These data demonstrated the very variable nature of pesticide residues in air, and it seems probable that residues in air are large only close to land areas which are frequently sprayed. However, Riseborough[214] suggested that insecticides can be transported large distances by global air currents, in the same way as occurs with radioisotopes such as strontium-90 and cesium-137, and then fall out on to land or water in a pattern dependent on local precipitation. He also suggested that dust deposits in glaciers, which can be accurately dated with lead-210, could be used to estimate fallout of chlorinated hydrocarbons because increased deposits of talc, which is used as a diluent for insecticides, can be correlated in time with the use of chlorinated hydrocarbon insecticides.

To confirm how widespread pesticide resi-

TABLE 3

RESIDUES OF PESTICIDES IN AIR & RAINWATER

RESIDUES (pp 10^{-12})*

Location	Reference	No. of sites	Medium	DDT & related compounds		γ BHC		Aldrin		Dieldrin		Chlordane		Heptachlor		Toxaphene	
				Max.	Mean	Max.	Mean	Max.	Mean	Max.	Mean	Max.	Mean	Max.	Mean	Max.	Mean
Great Britain	Abbott et al. 1966 (1)	1	Air	13.0	13.0	-	11.0	-	-	-	21.0	-	-	-	-	-	-
Great Britain	Tarrant & Tatton 1968. (249)	7 (4 dates)	Rainwater	79.3	79.3	230.0	60.3	-	-	35.0	7.6	-	-	-	-	-	-
Great Britain	Wheatley & Hardman 1965 (282)	1 (3 dates)	Rainwater	-	-	120.0	97.0	-	-	36.0	28.0	-	-	-	-	-	-
				1.0	1.0	164.0	100.0	-	-	25.0	20.0	-	-	-	-	-	-
				3.0	3.0	52.0	29.0	-	-	16.0	9.0	-	-	-	-	-	-
U.S.A.	Tabor 1966 (248)	6	Air (in agricultural communities)	0.003	0.004	-	-	0.004	0.0006	-	-	0.006	0.002	-	-	0.015	0.002
		1	Air (non-urban community)	4.2	4.2	-	-	-	-	-	-	0.03	0.016	-	-	-	-
		1 (3 dates)	Air (eastern city)	56.1	37.3	70.0	23.0	-	-	-	-	-	-	-	-	-	-
U.S.A.	Weibel et al. 1966 (279) (summarized in Cohen & Pinkerton 1966 (43)	3	Rainwater	102.0	210.0	70.0	23.0	-	-	-	-	-	-	-	-	-	-
			Dust in water	800,000.0	600,000	-	-	-	-	-	3,000	-	500,000	-	40,000	-	-
Barbados	Riseborough et al. 1968 (215)	1 (15 dates)	Dust in air	45,000.0	38,280	-	-	-	-	8,100	2,186	-	-	-	-	-	-

* ng/litre for water (nanograms/liter)
  µg/m3 for air (micrograms/cubic meter)

dues are in air, a large-scale monitoring program is necessary; indeed a global survey might well be combined with those already existing for measuring radionuclide levels. Available evidence indicates that the amounts of pesticides carried in the atmosphere are insufficient to account for the unexpected contamination of untreated soil and water that has sometimes been reported. Many pesticides are degraded by ultraviolet rays in sunlight, so it is possible that residues carried in the upper air may break down during transport.

The amounts of pesticides in air are unlikely to be harmful to human beings breathing the air. In a preliminary study by Barney[12] there was evidence that the amounts respired are very small by comparison with the amounts consumed with food. (Table 4, see below). Tabor[248] calculated that the average daily intake of pesticides by respiration by human beings in the United States was no more than between 2.0 and 32.0 μg. Current evidence indicates that hazards from pollution of the atmosphere by pesticides are still small, particularly compared with those from pollution of air by other materials. It is doubtful if amounts in the atmosphere will increase; on the contrary they may already be becoming less,[249] but, undoubtedly, further detailed studies of the distribution and fluctuations of pesticides in the atmosphere are needed.

Pesticides in Water

Pesticides can very easily reach ponds, lakes, or rivers by a variety of routes. They may be applied to water as aerial sprays to control pests such as mosquitoes,[280] blackflies,[7] or midges;[85] they may fall on water accidentally when applied as aerial treatments to control forest pests,[104 142] or agricultural pests;[199] they may reach water by surface runoff from soil;[84 295] they may be discharged with sewage effluents[131] or industrial effluents;[204] or they may be carried down from the atmosphere in rain.[249]

It is still difficult to assess the relative importance of sources of insecticide residues found in rivers, but Nicholson[204] stated that the two principal sources of contamination are runoff from agricultural land, and discharge of industrial wastes, either from industries that manufacture or formulate pesticides, or from those that use these compounds in their manufacturing processes, e.g., moth-proofing chemicals. The latter source of residues may be important locally but aerial spraying probably contributes more to the steady replenishment of residues in rivers, particularly in the United States. Runoff from agricultural land is a steady source of residues in rivers, but amounts from this cause are relatively small. Careless applications of insecticides, and disposal or washing of insecticide containers may add to residues in

TABLE 4

PESTICIDE INTAKE FROM AIR AND DIET IN UNITED STATES

Insecticide	Mean micrograms/kg body weight/day		
	From air	From diet	FAO/WHO acceptable daily intake
DDT & related compounds	0.227	0.8	10.0
Dieldrin	0.046	0.08	0.1
Endrin	0.01	0.004	-
Lindane	0.002	0.07	12.5

From Barney 1969 (12)

natural waters. Many of the main sources of contamination could be controlled, so that fewer pesticides reach fresh waters.

Middleton and Lichtenberg[187] reported insecticide residues from streams and rivers in the United States about ten years after the introduction of chlorinated hydrocarbon insecticides. It soon became obvious that many waterways in the United States were contaminated with these insecticides, and programs to monitor residues in major rivers were begun from 1962 onwards; the results of some of these surveys are summarized in Table 5 (see page 21). There has been no monitoring program in Great Britain, but Lowden et al.[179] surveyed several British rivers, although their survey was not comprehensive. In the United States, all of the different chlorinated hydrocarbon insecticides have been reported from rivers, sometimes in large quantities; Green et al.[106] reported that dieldrin was almost ubiquitous in rivers by 1964, with little consistency in its geographical distribution. By contrast, in Great Britain only DDT, BHC, and dieldrin have been found; this is not surprising because it corresponds with the usage of insecticides in that country; whereas in the United States other hydrocarbons such as chlordane, heptachlor, toxaphene, and endrin are used in large amounts. The amounts of residues reported have been quite variable, but although DDT has been found in the greatest quantities it is not always the most common residue. Endrin has been found in water in very large amounts in the United States.

Chlorinated hydrocarbon insecticides are not usually in solution in the water because they are all of very low solubility, and the residues reported in the surveys are of insecticides carried on particulate matter suspended in the water. When an insecticide reaches natural waters, a large proportion disappears rapidly.[36,234] For example, Wiedhass et al.[280] kept water containing DDT above soil covered with filter paper, and after six hours, 56% of the insecticide was in the soil; whereas, after 24 hours a further 22% had moved into the soil. When pesticides were monitored in the Mississippi River delta,[259] much larger quantities of chlorinated hydrocarbons were reported from the mud and sediment than from the water. In a study of the distribution of toxaphene in a lake

in New Mexico, Kallman et al.[149] found that after several days, there was a concentration of 0.01 to 0.28 ppm in the water, 0.04 to 0.13 in the bottom sediment, 0.4 to 18.3 in aquatic plants, and 2.5 to 15.2 in fish. Bridges et al.[24] followed the distribution of DDT in a pond and found that it was quickly concentrated into the bottom mud and vegetation. Lichtenstein et al.[175] found that aldrin and gamma BHC were much less stable in water if it contained lake mud.

Green et al.[106] considered that sedimentation was a major factor in removing chlorinated hydrocarbon insecticides from rivers in the United States and Keith[152] and Keith and Hunt[153] demonstrated that insecticides became partitioned between the water, suspended material, and bottom sediments. (Tables 6 and 7, see page 22). The solubility of an insecticide is an important factor, the more soluble chemicals taking longer to settle out. Nicholson et al.[206] showed that although DDT (which is very insoluble) was used in large amounts on soil in a watershed, it was not the most important residue in the water, but large quantities were found in the mud. The adsorption of lindane by lake bottom sediments was shown to be affected by the sediment suspension concentration, organic matter content, and clay content.

Thus insecticide residues are unlikely to occur in large quantities in standing water; only the turbulence of moving water can keep the particulate matter in suspension. In times of flood some of the deposited residues on the river bottom may be taken up into the main body of water again. It is clear from past surveys that the mud at the bottom of many rivers in the United States is heavily contaminated with pesticides, and will continue as a reservoir for periodic future contaminations of the water.

It was surprising that in the survey by Lowden et al.[179] insecticide residues in British rivers were greater than in the United States; Table 8 (see page 22) shows that the percentage of rivers with small concentrations of insecticides was much greater in the United States than in the north of England. Lowden et al.[179] could offer no explanation of this, other than that northern rivers were not typical of English rivers, but they estimated that in only 2% of their samples did the water contain sufficient insecticides to kill trout, although 30% contained about 10% of the lethal dose of insecti-

TABLE 5

PESTICIDE RESIDUES IN WATER SURVEYS
NANOGRAM PER LITRE (pp 10^{-12})

Location	Reference	No. of sites	Type of water	DDT & related compounds		Y BHC		Aldrin		Dieldrin		Chlordane		Heptachlor & H. epoxide		Endrin	
				Max.	Mean	Max.	Mean	Max.	Mean	Max.	Mean	Max.	Mean	Max.	Mean	Max.	Mean
U.S.A.	Breidenbach et al. 1967 (21)	99	Major river basins	149.0	8.2	4.0	t	-	-	68.0	6.9	-	-	155.0	6.3	116.0	2.41
U.S.A.	Anon. U.S.D.A. ARS 81-13 1966 (259)	10	Mississippi river delta	720.0	112.0	120.0	28.0	30.0	5.0	60.0	10.0	-	-	10.0	2.0	4230.0	541.0
U.S.A.	Green et al. 1967 (106)	109	Major rivers	127.0	8.3	56.0	2.2	-	-	167.0	5.9	75.0	0.1	19.0	0.1	69.0	3.6
U.S.A.	Warnick et al. 1966 (275)	48	Water areas (Utah)	32.4	9.7	-	-	-	-	-	-	-	-	-	-	-	-
U.S.A.	Keith & Hunt 1966 (153)	82	Water areas (California)	22.0	0.62	0.15	0.01	-	-	-	-	-	-	t	t	-	-
U.S.A.	Weaver et al. 1965 (277)	97	Major river basins	102.0	9.3	-	-	85.0	0.9	118.0	7.5	-	-	-	-	94.0	5.5
U.S.A.	Brown & Nishioka 1967 (26)	11	Western streams	145.0	10.3	20.0	2.8	5.0	0.2	15.0	2.3	-	-	105.0	2.6	40.0	1.4
Great Britain	Holden & Marsden 1966 (131)	-	Sewage effluents	130.0	36.0	-	-	-	-	300.0	200.0	-	-	-	-	-	-
Great Britain	Lowden et al. 1969 (179)	21	Sewage effluents	800.0	130.9	390.0	92.5	-	-	1900.0	145.0	-	-	-	-	-	-
Great Britain	Lowden et al. 1969 (179)	9	Yorkshire rivers	908.0	64.6	180.0	38.6	-	-	630.0	114.0	-	-	-	-	-	-
Great Britain	Lowden et al. 1969 (179)	7	British rivers	15.0	1.6	98.0	18.7	-	-	40.0	3.3	-	-	-	-	-	-

TABLE 6

DISTRIBUTION OF PESTICIDES IN A LAKE

Attributes	No. of samples	Average residues (ranges in parentheses)				
		DDT & related compounds	γ BHC	Toxaphene	Dieldrin	Heptachlor
Water (ppb)*	82	0.62 (0 - 22.0)	0.01 (0 - 0.15)	0.02 (0 - 0.32)	t	t
Particles (ppm)†	33	14.74 (1.8 - 78.0)	0	0	0	0
Bottom Sediment (ppm)†	39	4.44 (0.01 - 94.0)	t	0.03 (0 - 0.30)	t	t

From Keith & Hunt 1966 (153)

* μg/litre (microgram/litre)
† mg/kg (milligrams/kilogram)

TABLE 7

DISTRIBUTION OF PESTICIDES IN FOUR LAKES
TOTAL RESIDUES

Site	Average ppm DDT & related compounds		
	Suspended material*	Filtrate†	Total water sample†
Tule lake	6.0	0.0002	0.00045
Lower Klamath	9.0	0.0002	0.00035
Deer Flat	8.3	0.0001	0.0002
McNary	9.3	0.0001	0.0003

From Keith 1966 (152)

* mg/kg
† mg/litre

TABLE 8

PERCENTAGE OF SAMPLES WITH LESS THAN GIVEN CONCENTRATION OF INSECTICIDE FOUND DURING THE U.S. SURVEY OF MAJOR RIVER BASINS AND THE YORKSHIRE AND LANCASHIRE RIVER SURVEY

	Concentration of organo-chlorine insecticide (ng/l)							
	Dieldrin		DDT		TDE		γ BHC	
	10	100	10	100	10	100	10	100
U.S.A.	80	99	82	99	95	100	100	100
Yorkshire & Lancashire	62	87	73	95	71	87	41	96

From Lowden et al. 1969 (179)

cide. Their estimate did not take into account, however, the uptake of the insecticide into organisms which provide food for trout. It seems probable that much of the residues in northern English rivers may originate from industrial effluents. Although chlorinated hydrocarbon insecticides have been commonly reported from rivers, the amounts in drinking water supplies seem to be smaller than these. Nicholson[205] estimated the maximum allowable quantities of pesticides in drinking water for human beings. (Table 9, see below). It can be seen that these greatly exceed those commonly occurring in the water supplies, so this does not present a serious hazard.

The principal hazards of insecticide residues in water are twofold; large numbers of aquatic invertebrates and fish may be killed, or the residues may be taken up into the tissues of these organisms. It is clear that occasionally the amounts of pesticides in some rivers are sufficient to kill some fish, and this hazard may increase if residues in rivers continue to build up. There is evidence[128 143 199 203] that insecticides kill large numbers of aquatic invertebrates which provide food for fish, and help to keep the water nonstagnant by feeding on plankton and algae; fortunately, there are also good indications that when these organisms are killed by large, local applications of insecticides, there is usually a rapid repopulation. Although there have been some disastrous examples of aquatic organisms being killed by insecticides,[135 229] usually the area has become subsequently repopulated.

There is little information on the amounts of pesticides in sea water; indeed it is difficult to see how large amounts of pesticides could reach sea water. The most likely source would be through rain or dust in the atmosphere, or the outflow from major rivers into estuaries. The data quoted in the section on pesticides in air and rainwater show that the amounts that might reach the sea from the atmosphere are very small, and if diluted by the volume of water in the sea would be infinitesimal. Riseborough[215] calculated the input of pesticides into San Francisco Bay from San Joaquin River. Concentrations of residues in the San Joaquin River are of the order of 0.1 μg/liter and the mean outflow is 189×10^{12} liters/year which amounts to a total pesticide output of 1,900 kg/year into the Bay. A similar calculation for the Mississippi River estimates that 10,000 kg/year of pesticide residues pass out into the Gulf of Mexico. Even such large amounts di-

TABLE 9

SURFACE WATER CRITERIA FOR PESTICIDES IN PUBLIC WATER SUPPLIES (ng/l)

Pesticide	Permissible criteria	Desirable criteria
Aldrin	17	Absent
Chlordane	3	"
DDT	42	"
Dieldrin	17	"
Endrin	1	"
Heptachlor	18	"
Heptachlor epoxide	18	"
Lindane	56	"
Methoxychlor	35	"
Organic phosphates plus carbamates	100	"
Toxaphene	5	"
2, 4-D plus 2, 4, 5-T, plus 2, 4, 5-TP	100	"

From Nicholson, 1969 (205)

23

luted in the ocean would produce very small concentrations. Thus it is very difficult to account for the pesticide residues that have been reported in marine organisms.[214] [224] Butler[33] stated that the extent of pesticide pollution in estuaries of the North American continent is largely unknown, but fish and oysters from coastal areas contain pesticides. Riseborough[214] found more DDT in some marine fish than has been reported from fresh water fish living in contaminated streams.

Until we know more of the sources of pesticide residues in both fresh water and sea water, it is difficult to assess whether they will increase or decrease in the future. Nicholson[205] quoted data from his work on pesticide contamination from use of agricultural chemicals in a watershed which showed that, in years when pesticide usage was minimal, residues in water and mud fell dramatically. Many of the potential sources of these insecticides, particularly those from industrial discharges, could and should be controlled. In this context, contamination by pesticides is only a small part of the much larger problem of contamination by effluents from industrial areas. Although we know that pesticides disappear rapidly from water, we have little information on the rate at which chlorinated hydrocarbon insecticides break down in mud on river bottoms, and until this is available, future contaminations cannot be assessed. There is recent evidence that DDT[107] and BHC[202] break down faster in mud under anaerobic conditions, and it seems probable that much of the disappearance of insecticides from rivers is due to microorganisms, although codistillation and evaporation may be also important.

OCCURRENCE AND PERSISTENCE OF PESTICIDES IN THE BIOTA

Pesticides in the Soil Fauna and Flora

Residues of chlorinated hydrocarbon insecticides in the tissues of aquatic organisms were reported long before those in the bodies of invertebrates which inhabit soil, and the importance of these residues in animals which are usually in the lower trophic levels of food chains has been greatly neglected. In 1958, Barker[10] found large residues of DDT in earthworms, after DDT had been sprayed on elm trees to control insect vectors of Dutch Elm disease, and in 1962, Doane[65] confirmed that large residues occurred in the tissues of earthworms. Thereafter, several other workers discovered residues of DDT, BHC, and dieldrin in earthworms and, moreover, in concentrations considerably greater than those in the surrounding soil (Table 10, see page 25). Aldrin was found only in small quantities, but this is probably because it is metabolized to dieldrin in the earthworms. Other soil invertebrates, particularly slugs, were also found to concentrate pesticides in their tissues.[50] [57] [265] Presumably, many of the arthropods that live in or on the surface of soil also contain quantities of these hydrocarbon insecticides.[57] [203] Cramp and Conder[48] reported residues of 0.2 to 0.29 ppm of dieldrin in beetles, 0.06 in woodlice, 0.17 in cutworms, and up to 2.0 in earthworms from arable fields. Some slugs had as much as 10.3 ppm of endrin and 8.8 ppm of DDT.

Birds feed readily on invertebrates from soil, and the earthworms containing DDT[10] were probably responsible for the deaths of large numbers of American robins that fed upon them, although there was evidence that many insects taken as food, also contained DDT. There is little other evidence, other than circumstantial, that residues in soil invertebrates cause the death of birds, but it is reasonable to suppose that one source of chlorinated hydrocarbon insecticides found in the tissues of birds and in their eggs, may be from those invertebrates taken as food. Further surveys of residues in invertebrates, and the results of experimentally feeding these to birds would be of considerable value.[145] Wheatley and Hardman[283] investigated the relationship between amounts of pesticide residues taken up into earthworm tissues, and the amounts in the soil. They found that not all earthworm species concentrated pesticides to the same degree, and the largest concentrations were found in *A. chlorotica*. Smaller concentrations of residues of dieldrin, DDT, and gamma BHC occurred in the larger species (*L. terrestris, A. longa* and *O. cy-*

TABLE 10

RESIDUES OF PESTICIDES IN SOIL INVERTEBRATES AND THEIR ENVIRONMENT

RESIDUES IN (pp 10^{-6})*

Location	Reference	No. of sites	Source of residue	DDT & related compounds			Y BHC			Aldrin			Dieldrin		
				Max.	Mean	Con. factor†	Max.	Mean	Con. factor	Max.	Mean	Con. factor	Max.	Mean	Con. factor
Great Britain	Stringer & Pickard 1963 (247)	1	Soil (orchard)	26.6	11.4	-	-	-	-	-	-	-	-	-	-
			Earthworms	-	-	-	-	-	-	-	-	-	-	-	-
			(L. terrestris)	13.7	7.7	0.67	-	-	-	-	-	-	-	-	-
			Other sp.	14.0	8.1	0.71	0.01	t	-	0.7	0.15	-	0.7	0.15	-
Great Britain	Davis 1968 (55)	10	Soil (arable)	0.8	0.3	-	0.05	t	5.0	0.4	0.05	0.333	0.8	0.3	2.00
			Earthworms	0.05	t	0.06	0.05	t	5.0	0.4	0.05	0.333	0.8	0.3	2.00
Great Britain		2	Soil (orchard)	17.2	9.7	-	0.08	0.04	8.0	-	-	-	-	-	-
			Earthworms	40.0	19.6	2.10	0.3	0.1	2.5	-	-	-	-	-	-
Great Britain	Edwards (unpublished)	2	Soil (arable)	-	-	-	-	-	-	0.35	0.24	-	0.38	0.28	-
			Earthworms	-	-	-	-	-	-	0.11	0.09	0.374	1.6	1.3	4.64
Great Britain	Wheatley & Hardman 1968 (283)	1	Soil (arable)	-	0.93	-	-	0.004	-	-	0.72	-	-	0.64	-
			Earthworms	-	-	-	-	-	-	-	-	-	-	-	-
			L. terrestris	-	1.1	1.18	-	0.006	1.5	-	0.05	0.069	-	1.6	2.5
			A. longa	-	1.34	1.43	-	0.006	1.5	-	0.28	0.39	-	2.2	3.44
			A. caliginosa	-	2.5	2.68	-	0.011	2.75	-	0.52	0.72	-	3.8	5.9
			A. chlorotica	-	4.6	4.86	-	0.013	3.25	-	0.98	1.36	-	4.6	7.19
			A. rosea	-	2.6	2.78	-	0.017	4.25	-	0.64	0.88	-	3.9	6.09
			O. cyaneum	-	1.24	1.33	-	0.008	2.00	-	0.84	1.16	-	2.4	3.75
Great Britain	Davis & Harrison 1966 (57)	2	Soil	0.8	0.3	-	0.1	t	-	0.7	0.15	-	0.7	0.15	-
			Beetles	2.3	-	2.81	-	-	-	-	-	-	0.2	-	0.29
			Slugs	-	-	-	-	-	-	-	-	-	0.3	-	0.42
			Soil	17.2	9.7	-	0.08	0.04	0.1	-	-	-	-	-	-
			Beetles	5.2	-	0.31	-	-	-	-	-	-	-	-	-
			Slugs	40.1	-	2.33	-	-	-	-	-	-	-	-	-
U.S.A.	Barker 1958 (10)	1	Soil	-	9.3	-	-	-	-	-	-	-	-	-	-
			Earthworms	-	-	-	-	-	-	-	-	-	-	-	-
			L. terrestris	-	19.2	2.06	-	-	-	-	-	-	-	-	-
			L. rubellus	-	680.0	73.10	-	-	-	-	-	-	-	-	-
			O. lacteum	-	173.0	18.6	-	-	-	-	-	-	-	-	-
			H. zeteki	-	492.0	52.8	-	-	-	-	-	-	-	-	-
			H. caliginosis trapezoides	-	106.0	11.3	-	-	-	-	-	-	-	-	-
U.S.A.	Hunt 1965 (136)	1	Soil	-	9.9	-	-	-	-	-	-	-	-	-	-
U.S.A.	Hunt 1965 (136)		Earthworms	-	-	-	-	-	-	-	-	-	-	-	-
			L. terrestris	-	140.6	14.2	-	-	-	-	-	-	-	-	-
U.S.A.	In Dustman & Stickel 1966 (71)	3	Soil	3.0	1.8	-	-	-	-	-	-	-	-	-	-
			Earthworms	25.0	17.0	9.44	-	-	-	-	-	-	-	-	-
U.S.A.	U.S.D.I. 1967 (266)	67	Soil	-	0.98	-	-	-	-	-	-	-	-	0.006	-
			Earthworms	-	9.64	9.8	-	-	-	-	-	-	-	0.76	126.6
U.S.A.	U.S.D.I. 1966 (265)	1	Slugs	-	42.7	-	-	-	-	-	-	-	-	0.43	-
			Slugs	-	19.7	-	-	-	-	-	-	-	-	0.21	0.48
U.S.A.	U.S.D.I. 1967 (266)	2	Soil	-	-	-	-	-	-	-	-	-	0.6	0.4	-
			Earthworms	-	-	-	-	-	-	-	-	-	4.0	3.0	7.5
U.S.A.	Doane 1962 (65)	1	Soil**	-	12.5	-	-	-	-	-	-	-	-	-	-
			Earthworms	-	43.0	3.44	-	-	-	-	-	-	-	-	-
Great Britain	Cramp & Conder 1965 (48)	1	Soil	0.69	-	-	-	-	-	-	-	-	-	-	-
			Earthworms	1.20	-	1.73	-	-	-	-	-	-	-	-	-
			Slugs	2.55	-	3.70	-	-	-	-	-	-	-	-	-

* mg/kg.

† Concentration factor = $\dfrac{\text{concentration in animal}}{\text{concentration in soil.}}$

** ppm calculated from initial dose rate.

aneum) than in the smaller species (*A. caliginosa, A. chlorotica* and *A. rosea*); these amounts are probably correlated with the habits of the different species. The smaller species live and feed mainly in the upper few inches of soil where residues are most likely to be concentrated in both arable and orchard sites. In contrast *L. terrestris* lives in well-defined burrows, often four to six feet deep, and probably feeds mainly on surface debris. When large concentrations of residues have been found in surface debris, residues in *L. terrestris* have approximated those in other species.[247]

There was evidence from Wheatley and Hardman's[283] work (Figure 2, see page 26), that the relationship between the amount of residues in the soil and in the worms was not linear, but that there was proportionately less concentration into tissues from soils containing large quantities of residues (Figure 1). It is clear from Figure 2 that the amount of concentration would be expected to change from about five- to ten-fold when residues in the soil are between 0.001 and 0.01 ppm, to less than unity when concentrations exceed about 10 ppm in the soil. These results point out the necessity for information on different doses and on the ecology of species when uptake of residues is

FIGURE 2

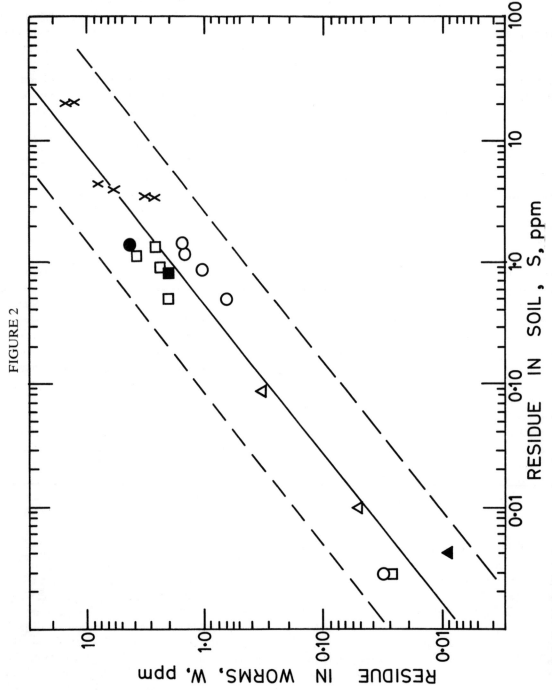

Relationship between the concentration of organochlorine insecticides in earthworms and associated soil. From Wheatley and Hardman, 1968.[283]

studied.

Stickel et al.[239] and Jefferies and Davis[145] studied the uptake of heptachlor and dieldrin, respectively, into earthworms, and then into the tissues of birds which were given contaminated earthworms as food. Stickel et al.[239] found that feeding worms containing about 0.65 ppm of heptachlor gave residues of 1.7 ppm in woodcocks, and Jefferies and Davis,[145] after feeding earthworms containing between 18.4 and 24.9 ppm to song thrushes, found 0.09 to 4.03 ppm in the birds. In neither experiment did pesticides become concentrated into the bird from the invertebrate.

Pesticide residues in soil may kill soil invertebrates, or may not kill them but be taken up into their tissues, or they may have sub-lethal effects. It is well established that there are residues in earthworms and slugs (Table 10, see page 25) and there is evidence[152 196 203] that many insects carry doses of insecticides that do not kill them; for instance, Newsom[203] quotes residues of chlorinated hydrocarbons as high as 12.9 ppm in carabid beetles, and Davis and French[56] found even larger residues in these insects. These invertebrates provide food for birds and mammals, and may be one of the sources of residues in higher organisms. Pesticide residues may also affect the invertebrate containing the residues,[196 203] the most important effect probably being on the reproductive potential of insects. There are numerous examples of this; for instance, Beard[13] showed that development of the ovaries of the house fly were suppressed by DDT, and the numbers of pink bollworm that mate have been lessened by DDT;[3] reduced egg fertility is very common. The behavior of insects is often affected, the usual result being that of hyper-activity. When exposed to sublethal doses of pesticides, many insects gradually become resistant and require larger doses to be killed.

It is extremely difficult to assess the hazards caused by pesticide residues in soil invertebrates. Although it can be expected that as fewer persistent insecticides are used and their residues in invertebrates will become less, there is evidence that invertebrates also take up some of the organophosphorus and carbamate insecticides that are gradually replacing the chlorinated hydrocarbons. These insecticides may be equally poisonous, although they are probably less persistent in the birds' tissues. Certainly more attention to residues of pesticides in invertebrates is necessary.

Pesticides in Aquatic Organisms

Chlorinated hydrocarbon insecticides are so toxic to aquatic invertebrates and fish that these organisms have been used to bioassay insecticides for many years.[95] There have been several reviews on the hazards of pesticides in water to fish, the best being that of Johnson,[147] who summarized the toxicity of pesticides to fish. A more complex problem than that of toxicity is the accumulation of pesticides in various aquatic organisms. When pesticides reach water, they are rapidly adsorbed by the bottom sediment, plankton, algae, aquatic invertebrates, aquatic vegetation, and fish. Some of these organisms may take up the pesticides directly from water flowing through their bodies, or they may accumulate them from their food. These kinds of accumulations, through different trophic levels in aquatic organisms,[24 149 152 153] have often been demonstrated experimentally.

There is extensive literature on residues of chlorinated hydrocarbon insecticides in fish, but space does not permit consideration of all these studies, so some representative examples of the levels of residues of different insecticides found in fish are given in Table 11 (see page 28). The amounts of residues rarely exceed a few parts per million; if much higher than this, it was usually when fish were caught shortly after large quantities of insecticides had reached the water from aerial spraying of either water or surrounding land.[135] The most common residues in fish are DDT and dieldrin (Table 11) although Eades[73] reported BHC and aldrin from salmon and trout, Hunt[134] found endrin in striped bass, and Keith[152] and Keith and Hunt[153] reported toxaphene from several species of fish. Heptachlor epoxide was reported in catfish and shad by the U.S.D.I.[262] Endrin, toxaphene, and heptachlor are used in the United States but not much in Great Britain, so it is interesting that they have not been found in fish in Great Britain. The insecticide most commonly found is DDT and this is also present in the largest quantities. The residues are not distributed equally in all the tissues of the fish but concentrated in its fatty tissues. Aquatic invertebrates take up chlorinated hy-

TABLE 11

INSECTICIDE RESIDUES IN AQUATIC VERTEBRATES
RESIDUES

Location	Fish	Amount of DDT ppm*	Amount of dieldrin ppm*	Reference
Great Britain	Brown trout	-	0.7 - 3.4	Holden 1966 (130)
" "	Trout (1963)	-	0.01 - 0.71	"
" "	Trout (1964)	-	0.06 - 0.92	"
" "	Perch	0.01 - 0.30	0.08 - 1.61	"
" "	Brown trout	0.10 - 0.26	0.01 - 0.03	"
" "	Trout	-	0.62	Eades 1966 (73)
" "	Salmon	0.71	0.28	"
" "	Sea trout	0.01 - 0.08	0.01 - 0.03	Holden 1966 (130)
" "	Plaice	0.023	0.038	Robinson et al. 1967 (224)
" "	Herring	0.080	0.057	"
" "	Sand eels	0.026	0.016	"
" "	Cod	0.012	0.009	"
" "	Whiting	0.021	0.040	"
" "	Porpoise	23.3	5.2	Holden & Marsden 1967 (132)
" "	Seal	8.3	0.25	"
Holland	Seal	16.3	1.3	Koeman & van Generen 1966 (157)
"	Sprats	0.10 - 0.15	0.16 - 0.42	Koeman et al. 1968 (158)
U.S.A.	Bluegill	1.29	-	Bridges et al. 1963 (24)
"	Bass	1.59	-	"
"	Bullhead	136.0	-	"
"	Bass	110.0	3.2	Keith & Hunt 1966 (153)
"	Black crappie	0.2	0.02	"
"	Pumpkin seed	0.06	0.01	"
"	Brown bullhead	0.19	-	"
"	Yellow perch	0.07	0.01	"
"	Tui chubb	0.11	0.01	"
"	Striped bass	64.0	0.8	Hunt 1964 (134)
"	Killifish	1.9	-	U.S.D.I. 1963 (262)
"	Gizzard shad	1.5	-	"
"	Brook trout	0.2 - 9.5	-	Cole et al. 1967 (44)
"	Sunfish	0.2	-	Mack et al. 1964 (181)
"	Bluegill	0.2	-	"
"	Carp	0.2	-	"
"	Rainbow trout	1.3	-	"
"	Rock bass	0.3	-	"
"	Smelt	0.4	-	"
"	White sucker	0.5	-	"
"	Alewife	0.7	-	"
"	Small mouthbass	0.7	-	"
"	Large mouthbass	0.9	-	"
Canada	Trout	0.03 - 0.16	-	Duffy & O'Connell 1968 (67)
"	Mackerel	0.01 - 2.75	-	"
"	Salmon	8.8 -30.0	-	"
U.S.A.	Pike	4.2	-	Mack et al. 1964 (181)
"	Yellow perch	0.3	-	"
"	Yellow perch	1.1	-	"
"	Brook trout	0.4	-	"
"	Brook trout	0.6	-	"
"	Lamprey	5.3	-	"
"	Channel catfish	3.8	-	"
"	Lake trout	3.6	-	"
"	Lake trout	6.2	-	"
"	Alewife	3.3	-	Hickey et al. 1966 (127)
"	Chub	4.19	-	"
"	Whitefish	5.05	-	"
"	7 spp. freshwater fish	5.7	-	U.S.D.I. 1966 (265)
"	Alewives	1.77 - 1.92	-	Hickey 1969 (127)
"	Northern anchovy	14.0	-	Riseborough 1969 (214)
"	Shiner perch	1.2	-	Harris 1964 (113)
"	English sole	0.5	-	Cohen et al. 1966 (43)
"	Jack mackerel	0.6	-	Robinson 1969 (220)
"	Hake	0.2	-	Gould 1966 (105)
"	White perch	4.1	-	Lyman et al. 1968 (180)
"	Pumpkin seed	3.0	-	"
"	Golden shiner	2.7	-	"
"	Yellow perch	3.1	-	"
"	Bluegill	2.6	-	"
"	Satin fin shiner	0.9	-	"
"	Common shiner	2.6	-	"
"	Alewife	2.3	-	"
"	Mummichog	6.3	-	"
"	Chain pickerel	1.6	-	"
"	Brown bullhead	2.4	-	"
"	European carp	4.9	-	"
"	White sucker	4.9	-	"
"	Killifish	8.2	-	"
"	Blacknose dace	0.8	-	"
"	Yellow belly sun fish	1.4	-	"
"	Fallfish	2.8	-	"
"	Rockbass	6.8	-	"
"	Creek chub	1.6	-	"
"	Small mouth bass	1.3	-	"
"	Crappie	21.1	-	"

* mg/kg

TABLE 12

Site	Reference	Organism	Insecticide	Amount		Concentration factor**
				In water (pp 10^{-9})*	In animals (pp 10^{-6})†	
U.S.A.	Terriere et al. 1965 (250)	Aquatic invertebrates	Toxaphene	0.63	1.43	2,270
U.S.A.	Butler 1965 (32)	Shrimp	DDT	0.5	0.14	2,800
U.S.A.	Cope 1966 (46)	Crayfish	DDT	20.0	1.47	73
U.S.A.	Bridges et al. 1963 (24)	Crayfish	DDT	20.0	2.0	100
U.S.A.	Cole et al. 1967 (44)	Crayfish	DDT	24.0	2.32	97
U.S.A.	U.S.D.I. 1963 (262)	Crayfish	DDT	20.0	0.33	16.5
U.S.A.	Hunt & Bischoff 1960 (135)	Plankton	DDT	20.0	5.0	250
U.S.A.	Keith 1964 (151)	Plankton	DDT	0.3	5.0	16,666
U.S.A.	U.S.D.I. 1964 (263)	Eastern oyster	DDT	10.0	151.0	15,100
		Eastern oyster	DDT	1.0	30.0	30,000
		Eastern oyster	DDT	0.1	7.0	70,000
		Pacific oyster	DDT	1.0	20.0	20,000
		Hand clam	DDT	1.0	3.0-9.0	3,000-9,000
		Eastern oyster	Dieldrin	1.0	3.5	3,500
		Commercial shrimp	DDT	0.5	0.14	280
		Sea squirt	DDT	100.0	20.0	200
				10.0	10.0	1,000
		Sea squirt	DDT	0.1	20.0	200,000
		Sea squirt	DDT	0.01	10.0	1,000,000
		Sea hare	DDT	0.01	1.78	178,000
		Crabs	DDT	50.0	7.2	144
		Snails	DDT	50.0	74.0	1,480
U.S.A.	Bugg et al. 1967 (29)	Oysters	BHC	0.33	0.006	18.2
			DDT	0.55	0.033	60.0
			Dieldrin	0.44	0.006	13.6
			Heptachlor	0.55	0.002	3.6
U.S.A.	Godsil & Johnson 1968 (103)	Clams	DDT	5.8	0.008	1.4
			Chlordane	6.6	0.006	0.9
			Endrin	10.5	0.013	1.2
U.S.A.	Loosanoff 1965 (177)	Common clam	DDT	0.1	7.0	70,000
U.S.A.		Oyster	Dieldrin	1.0	3.5	3,500
U.S.A.	Butler 1966 (33)	Hooked mussel	DDT	1.0	24.0	24,000
		Eastern oyster	DDT	1.0	26.0	26.000
		Pacific oyster	DDT	1.0	20.0	20,000
		European oyster	DDT	1.0	15.0	15,000
		Crested oyster	DDT	1.0	23.0	23,000
		Northern quahogs	DDT	1.0	6.0	6,000

* µg/litre

† mg/kg

** Concentration factor = $\dfrac{\text{concentration in animal}}{\text{concentration in water}}$

drocarbon insecticides from fresh and sea water, and some of the amounts reported are given in Table 12 (see page 29); some of these data, however, come from laboratory experiments where the animals were exposed to known concentrations of pesticides in water, so the degree of concentration of pesticides from water into the tissues of the invertebrates is also given in Table 12; these concentration factors range from a few times to many thousands. Fish can also concentrate pesticide residues into their tissues directly from water, probably mainly through the gills; Table 13 (see page 31) which summarizes the amounts in fish living in contaminated water, shows that the concentration factors are nearly as great as for those for aquatic invertebrates. This indicates that direct uptake of pesticides from water is a more important route for the pesticides into the bodies of fish than feeding on invertebrates which contain these chemicals.

The amounts of residues in the bodies of fish are mainly important because fish are a major source of food for human beings, other vertebrates, and birds. There is evidence that the amounts in fish become even more concentrated in the bodies of higher invertebrates which feed on them, but much of this is still circumstantial, and more experimental work following the passage of residues through food chains is required to ascertain how serious this potential concentration of persistent insecticides may be.

The amounts of persistent insecticides found in the bodies of fish are remarkably consistent, and it is usually possible to associate occasional larger amounts with nearby insecticide applications. However, there are clearly smaller residues in the tissues of marine fish than those from freshwater, and fish such as salmon, which migrate from freshwater to sea water, contain more pesticides than do other marine fish. Nevertheless, it is interesting that among the few data we have on residues of pesticides in seals and porpoises which feed on fish, there are considerably larger concentrations of residues in these animals than in fish (Table 11). More attention should be paid to the amounts of pesticides in the tissues of fish-eating mammals.

Most research on the effects of chlorinated hydrocarbons on fish has concerned toxicity, and there is relatively little work on sub-lethal effects. Cain[34] reported lowered resistance to disease and feeding rates below normal, and Cope[45] and Burdick et al.[29] described degeneration of reproduction. Other effects have included thickening of the gill membranes, lack of osmoregulation, lower blood counts, brain damage, and reduced body weights.[147] Until there is more systematic and regular monitoring of numbers of fish, it will be impossible to assess whether or not pesticides seriously deplete overall numbers of fish. Present evidence makes this appear unlikely, particularly because there seems to be little difference in susceptibility of different species to pesticides.

There is evidence that fish can develop resistance to chlorinated insecticides.[91] [154] Pesticide residues will continue to occur and fish continue to be killed until ways are found for lessening the amounts of pesticides which reach natural waters, or for extracting pesticides from them. It is possible that, with more careful use of pesticides and more rigorous government control of residues, there will be much less contamination of waters in the future.

Pesticides in Plants

Pesticides reach plants either as foliage sprays which are applied to control pests which feed on the leaves, or they are incorporated into the soil in which a crop is grown to kill soil inhabiting pests. When excessive amounts of a pesticide reach plants, they may damage the crop irrespective of whether the chemical is in the soil or on the plant foliage. Even when only small doses of insecticides are used, they can be taken up into the tissues of plants; some insecticides rely on this systemic action to kill pests, but fortunately, insecticides which are systemic are soon broken down to harmless compounds within the plant.

It is important that crops which are used for human or animal food should not contain residues of persistent pesticides, and in the United States, legal tolerances have been fixed for permissible amounts of residues that may occur in plant tissues that are to be used for food. In Great Britain there are recommendations for minimal times between insecticide applications and harvesting. As a result of these requirements, there have been very extensive studies of insecticide residues in plant material, but it

TABLE 13

<u>CONCENTRATION OF RESIDUES FROM WATER TO FISH</u>

Organism	Insecticide	Amount of residue		Concentration factor **	Reference
		In water (pp 10^{-9})*	In animal (pp 10^{-6})†		
Rainbow trout	DDT	20.0	4.15	207	Cope 1966 (46)
Black bullhead	DDT	20.0	3.11	155	"
Bluegill	Heptachlor	50.0	15.7	314	"
Catfish	Aldrin & dieldrin	0.044	0.07	1,590	Sparr et al. 1966 (236)
"	"	0.009	0.04	4,444	"
"	"	0.021	0.02	952	"
"	"	0.007	0.01	1,428	"
Buffabfish	"	0.023	0.09	3,913	"
"	"	0.007	0.21	30,000	"
Scaled sardine	DDT	0.1	0.11	1,100	Butler 1965 (32)
Rainbow trout	Toxaphene	0.41	7.72	18,829	Terriere et al. 1965 (250)
Fish	DDT	0.30	1.0 - 6.4	3,333 - 21,333	Keith 1964 (151)
"	Endrin	0.10	7.0	70,000	Langer 1964 (163)
"	Toxaphene	1.0 - 4.0	0.8 - 2.5	2,000 - 2,500	Kallman et al. 1962 (149)
" (5 spp.)	DDT	30 - 40	4 - 58	130 - 1,450	Crocker & Wilson 1965 (51)
Bullhead trout	DDT	20	2 - 4	100 - 200	Bridges et al. 1963 (24)
Fathead minnows	Endrin	0.015	0.15	10,000	Mount & Putnicki 1966 (198)
Croakers	DDT	0.1	2.0	20,000	Hansen 1966 (110)
Pinfish	DDT	1.0	12.0	12,000	"
"	DDT	0.1	4.0	40,000	"
Fish	Dieldrin & DDT	10.0	0.1 - 1.0	10 - 100	Holden & Marsden 1966 (131)
Trout	Dieldrin	2.3	7.7	3,300	Holden 1966 (130)
Chubs	DDT	5.8	0.029	5	Godsil & Johnson 1968 (103)
"	Chlordane	6.6	0.008	1.2	"
"	Endrin	10.5	0.050	4.7	"
Trout	DDT	20.0	4.0	200.0	U.S.D.I. 1963 (262)
Bluegills	Heptachlor	50.0	56.8	1,130.0	U.S.D.I. 1964 (263)
Fish	DDT	0.015	12.44	829,300	Mack et al. 1964 (181)
"	DDT	0.11	3.85	35,000	"
White catfish	DDD	14.0	30.4 - 129.0	2,172 - 9,214	Hunt & Bischoff 1960 (135)
Largemouth bass	DDD	14.0	19.7 - 25.0	1,407 - 1,785	"
Brown bullhead	DDD	14.0	15.5 - 24.8	1,107 - 1,771	"
Black crappie	DDD	14.0	5.4 - 115.0	386 - 8,214	"
Bluegill	DDD	14.0	6.6 - 10.0	471 - 714	"
Sacramento blackfish	DDD	14.0	10.9 - 17.6	778 - 1,257	"
Brook trout	DDT	24.0	17.3	710.0	Cole et al. 1967 (44)

* µg/liter

† mg/kg

** Concentration factor = $\dfrac{\text{Concentration in animal}}{\text{Concentration in water}}$

TABLE 14

INSECTICIDE RESIDUES IN SPRAYED OR DUSTED CROPS

Insecticide	Source of residue	Crop	Amount of insecticide (pp 10^{-6})*	Reference
DDT	-	Beans	0.3 - 1.5	Dormal et al. 1959 (66)
DDT	Dust	Beans	2.3	Brett & Bowery 1958 (22)
DDT	Spray	Tomatoes	0.5 - 5.1	Miller et al. 1957 (189)
DDT	Dust	Tomatoes	0.6	Brett & Bowery 1958 (22)
DDT	Spray	Lettuce	5.9 - 17.2	Waites & van Middelm 1958 (269)
DDT	Spray	Cherries	0 - 7.0	Maier-Bode 1961 (183)
DDT	Spray	Apples	0 - 12.4	Wolfe et al. 1959 (290)
DDT	Spray	Peaches	8.1 - 11.3	Dustan & Chisholm 1959 (70)
DDT	Spray	Strawberries	0 - 4.7	Fahey et al. 1962 (90)
DDT	Spray	Alfalfa	4.0 - 179.0	Gannon & Decker 1960(94a)
BHC	Dust	Beans	0.4	Brett & Bowery 1958 (22)
BHC	Dust	Tomatoes	0.3	"
BHC	Spray	Alfalfa	0.1 - 3.8	Fahey et al. 1962 (90)
Aldrin	Spray	Alfalfa	0.1 - 0.8	Gyrisco et al. 1959 (108)
Heptachlor	Spray	Alfalfa	0 - 13.3	Lichenstein & Medler 1958 (169)
Heptachlor	Spray	Alfalfa	1.0	Hardee et al. 1963 (111)
Endrin	Dust	Beans	0.5	Brett & Bowery 1958 (22)
Endrin	Dust	Tomatoes	0.3	"
Endrin	Spray	Black currants	0.4 - 4.9	Tew & Sillibourne 1961 (251)
Endrin	Spray	Alfalfa	2.0 - 3.7	Ely et al. 1957 (87)
Toxaphene	Dust	Beans	0 - 11.7	Wene 1958 (281)
Toxaphene	Dust	Beans	8.1	Brett & Bowery 1958 (22)
Toxaphene	Dust	Tomatoes	4.0	"
Toxaphene	Dust	Tomatoes	0.3 - 0.6	Wene 1958 (281)
Toxaphene	Dust	Lettuce	0.2 - 13.0	Wene 1958 (281)

* mg/kg

32

is outside the scope of this review to give a comprehensive survey of the amounts of residues that have been reported from plant tissues in such investigations. The amounts of insecticides reported in human food, including amounts in plant material, will be discussed in a later section of this paper. Typical insecticide residues that have been found in plant material have been reported and discussed by Marth[184] and by Finlayson and McCarthey.[92] Some of the typical amounts of residues of chlorinated hydrocarbon insecticides that have been found in plant material are summarized in Table 14 (see page 32). Not all these residues were actually in the plant tissues; some may have been merely on the cuticle, and these can be washed off before the plant is eaten.

Soon after the chlorinated hydrocarbon insecticides were first used in the soil, long-term studies to determine whether or not their residues in soil were phytotoxic to plants were made, as for example those by Morrison et al.,[197] Foster,[97] Allen et al.[4] and Dennis and Edwards.[60] Most of these studies showed that, with few exceptions, insecticide residues in soils did not decrease crop yields except when there were very large amounts; some insecticides in soil, in fact, stimulated growth. Thus it is unlikely that the residues that have now accumulated in soil (see earlier section) are directly hazardous to the growth of crops.

It was soon noticed that some of the chlorinated hydrocarbon insecticides, particularly BHC, and to a lesser extent aldrin, imparted slight off-flavors to root crops, but with the analytical techniques then available there were few reports of insecticide residues in the plant tissues. Later, as more sensitive analytical methods were developed, it became clear that small amounts of chlorinated hydrocarbon insecticides occurred in many plant tissues, particularly in those of root crops. It is not clear how chlorinated hydrocarbon insecticides move through plants; studies of their distribution in corn have shown that the lower parts of the plant contain larger amounts.[14] Some typical reports of residues in plant tissues, after uptake from soil are summarized in Table 15 (see page 34). There are many other similar reports both from Canada and the United States. Clearly, plants do not usually concentrate insecticides from soil into their tissues; in none of

the investigations listed in Table 15 was there a greater concentration of insecticide in the plant than in the soil in which it was grown. Nevertheless, although the concentrations of residues in the tissues are so small, they cannot be ignored, because when large amounts are consumed by man or vertebrates, these small amounts of residues may become concentrated in the tissues of the vertebrates. This is the main reason why there have been legal tolerances of the amounts of insecticides which may be consumed in food by human beings.

The relationships between the concentrations of insecticides in water, and those which occur in plankton, algae, and vascular plants which grow in the water are quite different to those between plants and soil. Table 15 shows that insecticides do become concentrated into the tissues of water plants, probably because these insecticides are very insoluble and have a greater affinity for biological material than for water. These aquatic plants and microorganisms provide food for fish and other aquatic organisms, and so may contribute to the concentration of chlorinated hydrocarbon insecticides in fish and, subsequently, higher invertebrates. Crops such as rice or cranberries,[61] which are grown with roots under water, take up much larger amounts of chlorinated hydrocarbon insecticides into their tissues than those grown in soil. Tsukano and Suzuki[254] showed that some crops could concentrate lindane from 4.0 ppm in water to 34.0 ppm in the stems and leaf sheaths, and 4.0 ppm in leaf blades ten days after treatment.

Walker[270] claimed to have demonstrated that plants took up residues of chlorinated hydrocarbon insecticides from the air. He placed kale, growing in pots, on plots which had been treated with insecticides, and when the foliage of the kale was analyzed, it contained between 0.04 and 0.2 ppm of residues.

The quantities of residues of chlorinated hydrocarbon insecticides in plant material should not be a serious hazard to vertebrates. There should be no problem in controlling residues in plant material that is intended for use as food, by establishing strict tolerance limits and a comprehensive monitoring scheme. Only if most soils became contaminated with these insecticides would there be a potential hazard, and this is unlikely to occur.

TABLE 15

MOVEMENT OF INSECTICIDES FROM SOIL OR WATER INTO PLANTS

Insecticide	Crop	Source of insecticide	Amount in source (pp 10^{-9})*	Residues Amount in plant (pp 10^{-6})†	Concentration or dilution factor**	Reference
Toxaphene	Aquatic plants	Water	0.41	0.21	512.0	Terriere et al. 1965 (250)
DDT	Aquatic vegetation	Water	200.0	75.0	375.0	Crocker & Wilson 1965 (51)
DDT	"	Water	20.0	31.0	1,550.0	Bridges et al. 1963 (24)
Organo-chlorines	Aquatic plants	Water	0.45	1.0	2,220.0	Keith 1966 (152)
Organo-chlorines	"	Water	0.35	1.1	3,171.0	"
Organo-chlorines	"	Water	0.23	0.8	3,478.0	"
Organo-chlorines	"	Water	0.30	30.3	100,000.0	"
DDT	Algae & moss	Water	0.33	0.01	33.0	Mack et al. 1964 (181)
DDT	Algae	Water	5.8	0.002	0.34	Godsil & Johnson 1968 (103)
Chlordane	"	Water	6.6	0.013	1.97	"
Endrin	"	Water	10.5	0.007	0.66	"
DDT	Vascular plants	Water	5.8	0.003	0.52	"
Chlordane	"	Water	6.6	0.003	0.45	"
Endrin	"	Water	10.5	0.006	0.57	"
			(pp 10^{-6})†	(pp 10^{-6})†		
Aldrin	Carrot roots	Muck soil	8.36	0.01(root)	0.0013	Hurtig & Harris 1966 (141)
Aldrin	"	Clay soil	0.48	0.01 "	0.021	"
Dieldrin	"	Muck soil	3.9	0.02	0.0051	"
Dieldrin	"	Clay soil	0.48	0.11	0.23	"
Heptachlor	Rutabagas roots	Soil	0.320	0.024	0.075	Saha & Stewart 1967 (232)
Heptachlor	Wheat foliage	Seed treated	543.0	0.015	0.036	Burrage & Saha 1967 (30)
Aldrin & dieldrin	"	Soil	1.8	0.014	0.0077	Saha & McDonald 1967 (231)
Aldrin & dieldrin	Cucumber fruit	Soil	3.7	0.113	0.031	Lichtenstein & Schulz 1965 (174)
Heptachlor	"	Soil	3.8	0.091	0.024	"
Dieldrin	"	Soil	1.4	0.043	0.031	"
Aldrin	Alfalfa	Soil	0.84	0.009	0.011	"
Heptachlor	"	Soil	0.78	0.028	0.036	"
Aldrin	Carrot roots	Soil	0.94	0.32	0.34	Lichtenstein et al. 1968 (176)
Aldrin	Potato tuber	Soil	0.94	0.07	0.074	"
Heptachlor	Carrot roots	Soil	0.49	0.36	0.73	"
Heptachlor	Potato tuber	Soil	0.49	0.05	0.10	"
DDT	Alfalfa foliage	Soil	1.39	0.113	0.08	Ware et al. 1968 (274)

TABLE 15 Continued

MOVEMENT OF INSECTICIDES FROM SOIL OR WATER INTO PLANTS

Insecticide	Crop	Source of insecticide	Amount in source (pp 10-6)†	Residues Amount in plant (pp 10-6)†	Concentration or dilution factor**	Reference
Dieldrin	Wheat	Soil	1.13	0.17	0.15	Wingo 1966 (288)
Dieldrin	"	Soil	18.39	1.07	0.04	"
Heptachlor	"	Soil	1.94	0.11	0.06	"
Heptachlor	"	Soil	9.74	0.44	0.05	"
Dieldrin	Legume foliage	Soil	0.05	0.0025	0.05	Saha et al. 1968 (230)
Aldrin	Peas	Sand	2.3	51.5(roots)	22.4	Lichtenstein et al. 1967 (168)
Aldrin	Peas	Sand	2.3	0.19(foliage)	0.082	"
Lindane	Peas	Sand	0.62	82.6(roots)	133.2	"
Lindane	Peas	Sand	0.62	22.9(foliage)	36.9	"
Aldrin	Carrot roots	Silt loam soil	1.34	0.53	0.4	Lichtenstein et al. 1965 (171)
Heptachlor	Carrot roots	Silt loam soil	1.33	0.98	0.7	"
Aldrin & dieldrin	Beet roots	Silt loam soil	2.3	0.13	0.06	Lichtenstein & Schulz 1965 (174)
Aldrin & dieldrin	Lettuce foliage	Silt loam soil	2.3	0.03	0.013	"
Aldrin & dieldrin	Turnips	Silt loam soil	2.3	0.05	0.022	"
Heptachlor	Beet roots	Silt loam soil	2.5	0.14	0.056	"
Heptachlor	Lettuce	Silt loam soil	2.5	0.04	0.016	"
Heptachlor	Turnips	Silt loam soil	2.5	0.09	0.036	"
Lindane	Carrot roots	Sandy loam soil	1.7	6.0	3.5	Lichtenstein 1959 (166)
Lindane	Potato tubers	Sandy loam soil	3.0	0.62	0.21	"
Lindane	Pea vines	Sandy loam soil	1.7	1.52	0.9	"
Lindane	Cabbage leaf	Sandy loam soil	1.7	0.37	0.22	"
DDT	Carrot roots	Sandy loam soil	24.8	3.17	0.13	"
DDT	Potato tubers	Sandy loam soil	24.8	1.63	0.07	"
DDT	Rutabagas	Sandy loam soil	24.8	0.9	0.04	"
Heptachlor	Soybeans	Soil	1.0	0.11	0.11	Bruce et al. 1966 (28)
Heptachlor	Oat seed	Soil	1.0	0.02	0.02	"

* µg/litre

† mg/kg

** Concentration or dilution factor = $\dfrac{\text{Concentration in plant}}{\text{Concentration in water or soil}}$

Pesticides in Vertebrates other than Fish

There have been very many reports during the last 15 years of residues of pesticides in mammals and birds and their eggs; more of the reports have concerned birds than any other groups or organisms. Tables 16, 17, 18 (see pages 37 - 40) summarize some of the amounts of insecticides that have been reported from birds and mammals in surveys; these tables are by no means comprehensive and quote only sufficient data to indicate the magnitude of residues that currently exist in the tissues of these animals. The amounts quoted may be somewhat biased toward larger estimates because, often, the carcasses of wild animals have been analyzed for insecticide residues after they have been found dead, in an attempt to determine the cause of death.

It is now clear that residues of chlorinated hydrocarbon insecticides occur in the tissues and eggs of many species of birds, both in Europe and North America,[219] and that these residues must have been accumulated from the food of the birds, whether this is plant or animal in origin. Insecticide residues have been reported from more than 118 species of birds in Great Britain,[48] and a census in the United States would probably show even more containing residues; in both countries there are few birds that have been analyzed which do not contain at least traces of chlorinated hydrocarbon insecticides. Attention was first focussed on the problem in Great Britain when large numbers of birds were found dead, associated with the use of insecticides and fungicides for seed-dressings, and it was shown experimentally by Turtle et al.[257] that the amounts of chlorinated hydrocarbon insecticides reported in the corpses of wild seed-eating birds found near agricultural land, were sufficient to have killed them. The United Kingdom Government placed restrictions on the use of seed-dressings, and subsequently allowed only seed planted in autumn to be dressed with persistent insecticides; heptachlor seed-dressings were banned completely. About the same time, there were occasional reports of large numbers of birds dying in circumstances which suggested chlorinated hydrocarbon insecticide poisoning as the cause of death.[10] [135] By 1962, the carcasses of sufficient random samples of birds had been analyzed to show that, although there were in-

secticide residues which consisted mainly of aldrin, dieldrin, or DDT, in the bodies of most birds, the concentrations differed greatly, and the quantities found were associated with the habits and food of the birds. Generally, there were much more insecticide residues in raptorial and fish-feeding birds than in herbivorous birds; particularly large amounts have been found in Herons and Great-Crested Grebes and their eggs.[272] (Tables 16 and 17). Residues of pesticides are rather similar for birds that have similar habits, even though they are from different countries; for instance, the bodies of herons in Great Britain contain about the same concentrations of residues as bald eagles and brown pelicans in the United States,[72] and the eggs of ospreys in the two countries contain similar concentrations of DDT.

The differences between the insecticide residues which have been found in birds from Great Britain and the United States are not so easily associated with the amounts of insecticides used in the two countries, as were the residues in soils, water, and invertebrates. For instance, heptachlor has been commonly found in British birds; Cramp et al.[49] found heptachlor residues in 38 species of British birds in 1966 and heptachlor has not been used since 1962, when it was used only as a seed dressing. Heptachlor residues were also found in larger amounts in bird-eating birds than in seed-eating ones, which suggests there is a concentration from prey to predator.[272] Endrin residues have also been reported from Great Britain[49] and it is only used there to control a mite attacking blackcurrants.

It is clear that some predatory birds are declining in numbers, although it is not yet certain that these declines are caused by insecticide residues. Bird-eating species such as Peregrine falcons and Sparrow hawks have decreased in numbers more than the fish-eating Herons and Grebes,[194] and Hickey[125] has reported similar decreases in raptors in the United States. One of the main problems is that populations of most wild vertebrates have not been assessed, so that changes occurring are not obvious.

Although large residues have been reported from birds, it is still not clear what the significance of these residues is, or even how these residues are transmitted from the physical envi-

TABLE 16

INSECTICIDE RESIDUES IN BIRDS

RESIDUES (pp 10^{-6})*

Location	Reference	No. of samples	Animal	DDT & related compounds Brain	DDT Liver	DDT Tissues	γ BHC Brain	γ BHC Liver	γ BHC Tissues	Dieldrin & aldrin Brain	Dieldrin Liver	Dieldrin Tissues	Heptachlor & H. epoxide Brain	Heptachlor Liver	Heptachlor Tissues
Holland	Koeman & van Genderen 1965 (157)	14	Birds of prey	8.9	3.3	18.1	4.7	8.1	5.4	3.0	10.0	7.5	0.9	0.6	0.6
		4	Owls	12.0	24.9	15.1	1.2	2.1	0.1	3.4	5.6	7.8	-	0.6	0.2
Great Britain	Robinson 1967 (219)	15	Birds of prey	-	8.0	-	-	1.6	-	-	1.8	-	-	2.1	-
		9	Owls	-	5.3	-	-	0.1	-	-	1.9	-	-	1.9	-
Great Britain	Walker et al. 1967 (272)	11-16	Kestrels	-	9.2	6.3	-	-	-	-	4.3	2.3	-	3.0	1.4
		8	Sparrow hawks	-	17.4	7.6	-	-	-	-	2.9	1.1	-	1.0	0.1
		12	Herons	-	19.4	7.1	-	-	-	-	7.8	3.3	-	-	-
		6	Moorhens	-	-	-	-	-	-	-	-	-	-	-	-
		6	Wood pigeon	-	-	-	-	-	-	-	-	-	-	-	-
		4	Thrushes	-	-	-	-	-	-	-	-	-	-	-	-
Great Britain	Robinson et al. 1967 (224)	2	Eider duck	-	0.25	-	-	-	-	-	0.12	0.1	-	-	-
		2	Herring gull	-	0.26	-	-	-	-	-	0.31	0.2	-	-	-
Ireland	Eades 1966 (73)	11	Pheasant	-	-	0.3	-	0.4	-	-	-	7.0	-	-	-
		3	Pigeon	-	-	0.12	-	0.05	-	-	-	3.1	-	-	-
		2	Thrush	-	-	-	-	-	-	-	-	-	-	-	-
Great Britain	Jefferies & Prestt 1966 (146)	3	Wood pigeon	-	-	0.06	-	-	0.15	-	-	3.0	-	-	0.4
			Mallard	-	-	5.1	-	-	0.05	-	-	-	-	-	-
Antarctica	George & Frear 1966 (100)	15	Skuas	-	0.24	0.16	-	-	-	-	-	-	-	-	-
Great Britain	Cramp et al. 1964 (50)	8	Grebes & herons	3.0	16.0	8.4	0.025	0.09	0.15	0.76	6.54	4.03	-	-	7.7
		13	Other water birds	-	-	0.18	0.008	-	0.002	-	-	0.059	-	-	0.88
		23	Insect feeding birds	-	-	0.79	-	-	0.016	-	-	0.69	-	-	0.003
Great Britain	Cramp & Conder 1965 (48)	20	Owls	3.7	5.36	2.17	-	0.44	-	1.6	2.35	1.2	0.38	0.38	0.38
		8	Birds of prey	3.4	3.11	1.42	-	0.003	0.022	2.05	2.06	0.56	1.2	1.73	0.16
		34	Hawks & falcons	-	7.5	3.2	-	-	0.17	-	0.9	1.51	-	0.24	1.55
		31	Owls	-	4.9	11.1	-	-	0.15	-	0.27	2.66	-	0.08	2.63
		9	Fresh water birds	-	-	2.84	-	-	0.15	-	-	5.5	-	-	0.02
		9	Marsh birds	-	-	0.3	-	-	0.02	-	-	0.24	-	-	0.04
		7	Marine birds	-	-	0.24	-	-	0.03	-	-	0.18	-	-	0.09
		8	Insect feeding birds	-	-	1.18	-	-	0.04	-	-	0.61	-	-	0.08

TABLE 16 Continued

Location	Reference	No. of samples	Animal	DDT & related compounds			γ BHC			Dieldrin & aldrin			Heptachlor & H. epoxide		
				Brain	Liver	Tissues	Brain	Liver	Tissues	Brain	Liver	Tissues	Brain	Liver	Tissues
Great Britain	Cramp & Conder 1965 (48)	12	Tree birds	-	-	0.41	-	-	0.03	-	-	0.16	-	-	0.37
Great Britain	Cramp & Olney 1966 (49)	115	Plant & insect feeding birds	-	-	55.6	-	-	0.23	-	-	4.7	-	-	1.86
		21	Plant feeding birds	-	-	1.55	-	-	2.58	-	-	2.82	-	-	0.44
		4	Marine birds	-	-	1.96	-	-	0.07	-	-	0.5	-	-	0.15
		52	Insect feeding birds	-	-	6.44	-	-	0.14	-	-	0.67	-	-	0.045
		19	Omnivorous birds	-	-	1.5	-	-	0.16	-	-	2.2	-	-	0.42
		10	Herons & grebes	-	-	25.2	-	-	0.34	-	-	11.6	-	-	0.07
		2	Marine birds	-	-	4.6	-	-	0.16	-	-	2.3	-	-	0.1
U.S.A.	Reichel et al. 1969 (212)	2	Bald eagle	77.5	165.4	175.8	-	-	-	4.7	10.5	5.9	0.04	0.05	0.04
U.S.A.	Turner 1965 (256)	69	Migrating birds	1.7	-	0.7	-	-	-	-	-	0.03	-	-	0.01
U.S.A.	Hickey et al. 1966 (127)	2	Long tailed ducks	-	-	6.3	-	-	-	-	-	-	-	-	-
		2	Ring billed gulls	7.1	-	28.0	-	-	-	-	-	-	-	-	-
U.S.A.	Stickel et al. 1968 (241)	12	Herring gulls	20.8	2.3	98.8	1.1	0.01	0.01	9.35	13.5	4.2	-	-	-
		5	Meadow lark	2.0	-	0.6	-	-	-	10.3	-	-	-	0.04	-
		7	Robin	1.9	-	-	-	-	-	12.5	-	-	-	-	-
		2	Starling	1.4	-	-	-	-	-	7.6	-	-	-	-	-
		2	Woodcock	0.7	-	-	-	-	-	-	-	-	-	-	-
U.S.A.	U.S.D.I. 1965 (264)	27	Grackle	-	-	3.3	-	-	-	-	-	3.3	-	-	0.3
		20	Robins	-	-	6.1	-	-	-	-	-	6.8	-	-	0.2
		24	Starlings	-	-	2.1	-	-	-	-	-	5.6	-	-	0.4
		6	Crows	-	-	46.1	-	-	-	-	-	-	-	-	-
U.S.A.	U.S.D.I. 1963 (262)	8	Sparrow	-	-	6.8	-	-	-	-	-	3.3	-	-	3.8
		3	Thrush	-	-	10.5	-	-	-	-	-	1.2	-	-	-
		9	Warbler	-	-	11.4	-	-	-	-	-	-	-	-	-
		2	Woodpecker	-	-	0.3	-	-	-	-	-	-	-	-	2.0
		3	Robin	-	-	0.2	-	-	-	-	-	-	-	-	2.0
		13	Ducks	-	-	-	-	-	-	-	-	-	-	-	-
		8	Ducks	-	-	-	-	-	-	-	-	-	-	-	-
U.S.A.	U.S.D.I. 1967 (266)	-	Pelicans	-	-	194.0	-	-	-	-	-	10.0	-	-	-
U.S.A.	U.S.D.I. 1965 (264)	42	Water fowl	-	-	0.73	-	-	-	-	-	-	-	-	-
		11	Loons	3.3	2.7	2.8	-	-	-	-	-	-	-	-	-

* mg/kg

38

TABLE 17

RESIDUES (pp 10^{-6})*

Location	Reference	No. of samples	Animal	DDT & related compounds	γ BHC	Dieldrin & aldrin	Heptachlor & H. epoxide
Great Britain	Ratcliffe 1965 (210)	11	Eggs (Rook)	0.2	0.1	0.2	t
		11	" (Magpie)	0.2	t	0.1	t
		11	" (Carrion crow)	0.4	0.1	0.1	t
		13	" "	0.4	0.1	0.3	0.1
		10	" (Raven)	1.1	0.3	0.8	t
		6	" (Buzzard)	0.5	t	2.0	t
		2	" (Merlin)	5.6	0.1	0.4	0.2
		7	" (Kestrel)	0.7	0.1	0.2	t
Great Britain	Robinson 1967 (220)	-	" (Peregrine falcon)	5.0	0.03	0.4	0.3
				13.7	0.1	0.6	0.8
		77	" (Hawks & falcons)	3.3	0.04	1.0	0.1
Great Britain	Robinson et al.[†] 1967 (224)	72	Shag	1.9	-	1.4	-
		34	Kittiwake	0.25	-	0.12	-
		14	Sandwich tern	0.69	-	0.24	-
		8	Guillemot	1.59	-	0.35	-
		6	Puffin	0.28	-	0.31	-
Great Britain	Moore 1965 (191)	10	Raven	1.0	0.3	0.8	0.02
Great Britain	Walker et al. 1967 (272)	13	Eggs (Buzzard)	0.6	-	0.8	-
		28	" (Golden eagle)	0.8	-	1.1	-
		8	" (Kestrel)	0.5	-	0.15	-
		24	" (Terns)	1.0	-	0.25	-
		16	" (Sparrow hawk)	9.0	-	1.55	-
		14	" (Peregrine falcon)	11.2	-	0.9	-
		5	" (Heron)	8.3	-	5.6	-
		8	" (Grebe)	5.9	-	0.4	-
		6	" (Herring gull)	0.5	-	0.3	-
		4	" (Guillemot)	2.9	-	0.6	-
		7	" (Pheasant)	0.03	-	0.05	-
		2	" (Moorhen)	0.2	-	0.1	-
		25	" (Carrion crow)	0.3	-	0.1	-
U.S.A.	Reichel & Addy 1968 (211)	94	" (Black duck)	4.4	-	0.14	-
U.S.A.	Stickel et al. 1966 (242)	2	" (Bald eagle)	19.0	-	0.7	-
U.S.A.	U.S.D.I. 1964 (263)	48	" (Black duck)	0.36	-	-	-
U.S.A.	Ames 1966 (5)	2	" (Osprey)	5.5	-	-	-
U.S.A.	Keith 1966 (152)	4	" (Herring gull 1963)	127.0	-	-	-
		9	" " 1964)	227.0	-	-	-
U.S.A.	U.S.D.I. 1965 (264)	12	Osprey	7.1	-	-	-
U.S.A.	U.S.D.I. 1966 (265)	-	Osprey	3.4	-	-	-
U.S.A.	Riseborough 1969 (214)	17	Brandts cormorant	0.3	-	-	-
		6	Murre	1.9	-	-	-
		7	Caspian tern	1.3	-	-	-
		3	Western gull	0.3	-	-	-

* mg/kg
† These involve means of data from other workers.

TABLE 18

INSECTICIDE RESIDUES IN TERRESTRIAL VERTEBRATES

RESIDUES (pp 10^{-6})*

Location	Reference	No. of samples	Animal	DDT & related compounds Brain	Liver	Tissues	Y BHC Brain	Liver	Tissues	Dieldrin & aldrin Brain	Liver	Tissues	Heptachlor & H. epoxide Brain	Liver	Tissues
Antarctica	Sladen et al. 1966 (235)		Penguins	-	0.039	0.077	-	-	-	-	-	-	-	-	-
Ireland	Bades 1966 (73)	1	Foxhound	-	-	0.2	-	-	0.54	-	-	0.48	-	-	0.4
Great Britain	Cramp et al. 1965 (48)	7	Rodents	-	-	0.08	-	-	0.06	-	-	0.61	-	-	-
Great Britain	Cramp et al. 1966 (49)	1	Red squirrel	-	-	-	-	-	0.05	-	-	-	-	-	0.06
		3	Long-tailed field mouse	-	-	0.01	-	-	0.01	-	-	-	-	-	0.01
		1	Fox	-	-	0.3	-	-	-	-	-	-	-	-	-
		2	Stoat	-	-	0.24	-	-	-	-	-	-	-	-	-
		1	Weasel	-	-	0.13	-	-	-	-	-	0.7	-	-	-
Holland	van Klingeren et al. 1966 (155)	25	Hares	-	-	-	-	-	-	0.1	8.4	-	-	-	-
U.S.A.	U.S.D.I. 1965 (264)	6	Rabbits	-	-	0.6	-	-	-	-	-	-	-	-	-
U.S.A.	Stickel et al. 1969 (241)	5	Cotton rat	1.1	1.2	0.05	0.25	0.5	0.01	8.2	24.1	5.0	0.3	0.7	0.03
		5	Cottontail rabbit	0.2	0.1	0.06	0.14	0.01	0.01	14.8	54.9	7.7	0.01	0.03	0.02
U.S.A.	Pillmore & Finley 1963 (209)	10	Mule deer	-	2.3	10.9	-	-	-	-	-	-	-	-	-
		24	Elk	-	-	7.1	-	-	-	-	-	-	-	-	-
		9	Deer	-	-	6.3	-	-	-	-	-	-	-	-	-
		7	Elk	-	-	12.6	-	-	-	-	-	-	-	-	-
U.S.A.	U.S.D.I. 1963 (262)	7	Mouse	-	-	-	-	-	-	-	3.2	-	-	-	6.4
		2	Rat	-	-	-	-	-	-	-	-	-	-	-	3.0
U.S.A.	Walker et al. 1965 (273)	-	Bear	-	-	0.06	-	-	-	-	-	-	-	-	-
		-	Goat	-	-	0.02	-	-	-	-	-	-	-	-	-
		-	Antelope	-	-	0.10	-	-	-	-	-	-	-	-	-
		-	Moose	-	-	0.10	-	-	-	-	-	-	-	-	-
U.S.A.	DeWitt et al. 1960 (63)	5	Cotton rat	-	-	-	-	-	-	-	-	-	-	-	5.9
		1	Field mouse	-	-	-	-	-	-	-	-	-	-	-	33.5
		1	Swamp rabbit	-	-	-	-	-	-	-	-	-	-	-	3.3
		1	White footed mouse	-	-	71.6	-	-	-	-	-	-	-	-	-
		2	Cottontail rabbit	-	-	1.8	-	-	-	-	-	-	-	-	-

* mg/kg

40

ronment into the higher levels of food chains. It is clear that when a large area is sprayed with a large dose of insecticide, residues occur subsequently in many parts of the contaminated ecosystem.[127] [135] [256] Although residues are largest in the upper trophic levels of food chains, it is not certain by which route they reach these animals. Several workers have studied the movement of residues through experimental food chains,[145] [239] but often the amounts in the higher trophic levels have been smaller than expected.

Robinson et al.[224] investigated the amounts of residues in food chains of marine organisms. They found that although the concentrations in the different trophic levels usually tended to increase (Figure 3, see page 42), it did not occur in each particular food chain. For example, the concentration of dieldrin was less in cod than in its main food, the sand eel, and the amount of dieldrin in plankton crustacea was greater than in any of the fish which feed on these organisms. The simple food chain of microplankton to mussels to eider ducks showed concentration of dieldrin at each stage. Thus, although it is established that amounts of pesticides often increase through food chains, the amounts are very variable, and much more research is required to establish the frequency and importance of this concentration.

Robinson et al.[224] suggested that the concentration of insecticide residues from the tissues of lower organisms at the bottom of food chains into those of vertebrates at the top of food chains may have more than one explanation. It is usually assumed that vertebrates can concentrate these residues from food into their tissues, but an alternative explanation can be made, if it is assumed that the amounts of residues in birds result from an equilibrium between the rate of intake and excretion of the insecticide, rather than a continuous accumulation. For example, if the vertebrate is less efficient in excreting chlorinated hydrocarbons than invertebrates on which it feeds, then the same sorts of accumulations would occur. Robinson et al.[224] considered there is evidence that both mechanisms may operate, and that further research is required to determine their different contributions to accumulations and residues.

Insecticide residues in birds can be either directly harmful to birds and perhaps kill them, or may harm other animals which feed on the contaminated birds. It has frequently been assumed that when dead birds contain insecticide residues, then the insecticide must have caused their deaths. Alternatively, declines in numbers of certain species of birds have been correlated or associated with the usage of insecticides and the occurrence of insecticide residues in the bodies of the birds or their eggs. Robinson[219] criticized such circumstantial evidence on the basis that analytical techniques were often imperfect and suggested that the amounts found in dead birds should be carefully compared with the amounts found in birds that have been experimentally poisoned with insecticides. Stickel[237] stated that in assessing hazards of chlorinated hydrocarbon insecticides to birds, three types of research effort seem essential. Firstly, experimental work should establish toxic limits and sublethal effects for chlorinated hydrocarbons and heavy metals separately, and variations in effects, due to physiological and environmental stresses, should be established. Secondly, quantities of insecticide residues in the environment should be determined in some form of monitoring program, with particular attention to amounts in the environments of species which are declining in numbers. Thirdly, the rates and routes of gain and loss, and change of residues in and among the various living and non-living components, are required to identify where hazards lie. These aims are very logical, but, in the past, too many studies have done no more than assess residues in a few examples of some isolated component of the environment as, for instance, several randomly collected corpses.

In recent years, there have been many studies of the effects of feeding birds on diets containing insecticides[8] [49] [145] [239] [240] [242] [257] [265] and these have shown that such diets can kill birds, and the studies established the amounts of residues in birds that have died. The results of such tests are not all clear-cut, however, because, commonly, the survivors of such diets contain more insecticide residues than the ones that died; when Stickel and Stickel[240] fed cowbirds on a diet containing DDT this occurred. Such diets usually contain much more insecticide than would occur in the natural diet, and often contain enough to cause acute toxicity,[240] [265] so that birds accumulate much more

FIGURE 3

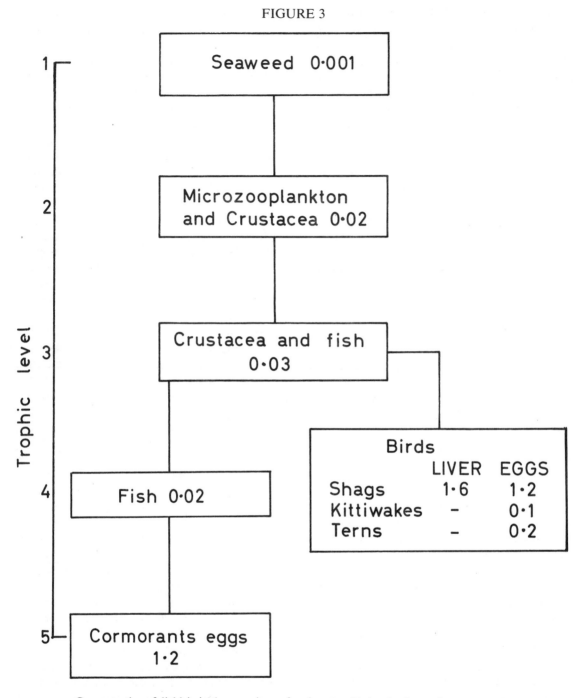

Concentration of dieldrin in the organisms of various trophic levels of a marine ecosystem. From Robinson, 1968.[218]

insecticides in their tissues (often up to 300 ppm) than would be found in wild birds that may have died from insecticide poisoning. There is some evidence that the amounts in the diet are proportional to the amounts found in tissues. This relationship holds only up to certain amounts, however, and it has been shown that, as in other animals, an equilibrium level is usually reached, when no further residues are accumulated, no matter how much insecticide is added to the diet.[52]

The variability in experimental results may be due to the mode of action of the insecticide; it has been shown by Stickel et al.[238] that healthy woodcocks could survive massive doses of DDT in their diet, but when they were starved, they rapidly succumbed, and this was ascribed to the insecticide being released into the circulation of the animals when reserve fat was metabolized. Bernard[15] showed that house sparrows fed on a diet containing DDT did not show symptoms of poisoning until starved, presumably due to mobilization of fat containing DDT. Moore[191] has suggested that more than 10 ppm of dieldrin and more than 30 ppm of DDT, in tissues other than fat, may be lethal to birds, but obviously this varies considerably between species and individual birds. DeWitt et al.[63] believed that residues of 20 to 30 ppm of DDT, 20 ppm of chlordane, or 6 to 29 ppm of heptachlor epoxide in the tissues of birds were evidence that birds died from these residues. Stickel[237] believed that the amounts of residues that are easiest to diagnose as a potentially lethal dose are the amounts of residues in the brain. The amounts in the brains of both birds and mammals that are believed to have died from pesticide poisoning are remarkably similar.

There is little experimental evidence that chlorinated insecticide residues affect the behavior of birds, but David[58] believed that the learning ability of bobwhite quail was impaired by 20 to 100 ppm of DDT in their diet. Birds' eggs also commonly contain residues of chlorinated hydrocarbon insecticides (Table 17); in Great Britain, Moore[193] reported that all eggs of sea birds analyzed had residues of between 0.4 and 3.5 ppm, and he thought that the consistency of these residues would make monitoring the amounts of residues in the eggs of seabirds a good index of the amounts of residues

in the environment. The results of a small-scale monitoring scheme, based on this idea, showed no major change in residues in marine birds' eggs between 1962 and 1967, but there were indications of a small decline in amounts of residues of DDT and dieldrin. Robinson et al.[224] claimed that there are seasonal variations in the amounts of residues in seabirds' eggs even within a single species in a restricted area, so perhaps such a scheme is not feasible. It is hard to see how seabirds acquire such insecticide residues, but the eggs of seabirds in Great Britain do contain much smaller residues than those of birds which live on freshwater fish.[195] Butcher[31] reported that the eggs of seven species of Australian birds contained DDT (0.6 to 12.0 ppm) and dieldrin (traces to 0.02 ppm). Hickey[125] associated amounts of chlorinated hydrocarbon insecticides, in the eggs of raptors, with reproductive failures, particularly the laying of thin shelled eggs which were subsequently eaten by the parents. Prestt[208] also reported a greater incidence of breakage and lower shell weight among eggs which contained residues and which were laid by birds containing large residues. Hickey and Anderson[126] correlated progressively thinner egg shells, over the past 25 years, with greater breakages, reduced hatching and insecticide residues, but there is some disagreement as to the validity of this conclusion.

There is some evidence that chlorinated hydrocarbon insecticide residues in their tissues lower the viability of chicks as well as eggs; for instance, De Witt[62] claimed to have shown that 1.0 ppm of dieldrin in the feed of quails reduced the viability of the chicks. Conversely, Azevedo[8] found that residues had no effect on either egg production or viability of pheasants. Moore[195] believed that insecticide residues affected the viability of young birds more than it did hatchability of eggs and size of clutches. Stickel[237] listed several investigations in which large insecticide residues in eggs did not seem to affect numbers which hatched at all. It seems that residues do affect bird reproduction, but it is difficult on present evidence to assess how important this is. It is also difficult to assess whether residues of chlorinated hydrocarbons will increase in the bodies of birds in the future, and, if they do, whether this will be of great importance. Current indications in Great

Britain are that residues are tending to decrease, and as compounds, other than the persistent insecticides, are used more and more in the United States, they may be expected to decrease there also. It is certain that birds lose residues when they are no longer exposed to insecticides.[15] [117] [162] [238]

Thus it seems it is proven that chlorinated hydrocarbon insecticides are a hazard to bird populations, but the seriousness of the hazard still remains to be assessed. There is, as yet, no experimental or conclusive evidence that the survival of any species has been threatened by the use of insecticides, but some species may have been excluded from certain areas.

The amounts of residues in terrestrial mammals have not been studied nearly so much, either in Great Britain or in the United States, as those in birds, but such data that we have, indicate that most mammals contain smaller concentrations of chlorinated insecticide residues than birds.[191] As in other animals, the principal residues are of DDT or dieldrin, but these are usually quite small although occurring commonly; Walker et al.[273] reported that amounts of DDT in game from regions untreated with insecticides were less than 0.1 ppm.

There is little evidence that predatory mammals contain larger concentrations of residues than herbivorous or omnivorous ones, but this may be because of insufficient samples and analyses.

There is, at present, no evidence of gross effects on these animals,[209] but much more information is required to assess the effects of continuous exposure to small doses. Extrapolating from laboratory tests with mice and rats, there seems to be no evidence of possible ill effects from the residues in wild animals. Neither do the amounts of residues in domestic animals have any apparent adverse effects on growth or reproduction.

Table 19 (see page 45) summarizes results from some experiments which have investigated the concentration of residues from food into the tissues of vertebrates. Some mammals can concentrate insecticides greatly into their tissues, others hardly at all. Robinson and Roberts[225] studied the accumulation, distribution, and elimination of organochlorine insecticides by vertebrates, and concluded that: firstly, the concentration of a particular chlorinated hydrocarbon insecticide in any vertebrate tissue was related to that in other tissues; secondly, that when the pesticide reached the animal in its food, the concentrations of the insecticides in the various tissues of the animal were a function of the concentrations in the diet, but that chronic oral ingestion of insecticides did not result in continuous rectilinear increase in the concentrations in their body tissues; the relation between time of exposure and amounts in the tissues was asymptotic in those experiments where sufficient data were available. They showed that, after exposure to insecticides had been terminated, the concentrations in the body tissues decreased in an exponential manner as stated by Robinson[221] who developed a compartmental mathematical model for the decrease. Williams et al.[287] showed that residues of five insecticides in cattle reached an equilibrium after eight weeks, and then declined after dosage was discontinued. Possibly, mammals can develop resistance to residues; for instance, Webb and Horsfall[278] have shown that pine mice in Virginia orchards have become resistant to endrin. There are indications that mice may produce fewer young when fed on a diet containing 200 to 300 ppm of DDT.[16] Dieldrin in the diet of white-tailed deer, at a rate of 25 ppm, decreased the reproduction rate of mature females.[159] There is little evidence that chlorinated hydrocarbon residues in the tissues of mammals significantly affect their behavior.

It seems that there is not much evidence that chlorinated hydrocarbon insecticides are hazardous to wild mammals, but more extensive monitoring of the amounts of residues in these animals is needed.

Pesticide Residues in Food and Feed

The residues of chlorinated hydrocarbon insecticides in the tissues of animals and plants have been outlined and discussed in earlier sections of this paper. These may occur in domestic animals, crop plants grown for human food, or in plants grown as feed for domestic animals. Milk and dairy products are particularly susceptible to accumulation of insecticides,[25] and for some years there has been strict supervision of the amounts of chlorinated hydrocarbon insecticides that are permitted in milk. In the United States there is a zero tolerance (based

TABLE 19

MOVEMENT OF INSECTICIDES INTO ANIMAL TISSUES FROM DIET

Location	Insecticide	Animal	Amount in source (pp 10^{-6})*	Amount in animal (pp 10^{-6})†	Concentration or dilution factor**	Reference
Great Britain	BHC	Pigeon	72.0	144.0	2.0	Turtle et al. 1963 (257)
Great Britain	Dieldrin	Pheasant	50.0	2.7 (muscle)	0.05	Cramp et al. 1966 (49)
Great Britain	Dieldrin	Song thrush	0.32-5.69	0.09-4.03	0.3-0.7	Jefferies & Davis 1968 (145)
U.S.A.	DDT	Pheasant	10.0	29.1	2.91	Azevedo et al. 1965 (8)
U.S.A.	DDT	Cowbird	500.0	74-171.0	0.1-0.3	Stickel & Stickel 1969 (240)
U.S.A.	Heptachlor	Woodcock	0.65	1.7	2.6	Stickel et al. 1965 (239)
			2.86	13.0	4.5	
U.S.A.	DDT	Bald eagle	5.0	0.7 (brain)	0.1	Stickel et al. 1966 (242)
				1.9 (liver)	0.4	
				35.7 (fat)	7.1	
U.S.A.	DDT	Bald eagle	800.0	291.0 (muscle)	0.4	U.S.D.I. 1966 (265)
				80.0 (brain)	0.1	
Great Britain	Dieldrin	Rat	10.0	15.85 (fat)	1.6	Robinson et al. 1969 (227)
				0.24 (brain)	0.02	
				0.85 (liver)	0.08	
				0.075 (blood)	0.007	
U.S.A.	Dieldrin	Rat	1.0	15.1	15.1	Street 1965 (243)
			10.0	67.5	6.7	
		Hen	0.25	10.2	40.8	
			0.75	35.7	47.6	
		Steer	0.25	0.8	3.2	
			0.75	3.5	4.7	
			2.25	8.7	3.9	
		Hog	0.25	0.4	1.6	
			0.75	2.8	3.7	
			2.25	4.3	1.9	
		Lamb	0.25	0.4	1.6	
			0.75	0.6	0.8	
			2.25	1.7	0.7	
U.S.A.	Dieldrin	White tailed deer	5.0	0.2 (brain)	0.04	U.S.D.I. 1967 (266)
			15.0	0.4 (brain)	0.03	
			45.0	2.5 (brain)	0.06	

* mg/l or mg/kg

† mg/kg

** Concentration factor = $\dfrac{\text{Amount in animal}}{\text{Amount in source}}$

45

on the sensitivity of analytical techniques available) for insecticide residues in milk. In contrast to residues in milk, the control of residues in other animal-derived foods is not very great; this is presumably because of the very great variability in amounts of pesticides in the tissues of animals, and the difficulties of monitoring the amounts of chemicals in such animal tissues. Residues of chlorinated hydrocarbon insecticides vary so much in animal tissues because they reach the animal indirectly, usually when it eats insecticide-contaminated food, and unless the residues in the animal's food are strictly controlled, insecticides will be taken into the body tissues of the animal. It is extremely difficult to control the diet of domestic animals, hence their frequent exposure to insecticides. Pesticides in animal tissues must be an important source of residues in human body fat; Hayes et al.[123] found that the body fat from people abstaining from eating meat contained about half as much DDT as fat from people in the general population.

It is easier to control and limit the residues of insecticides in vegetables and other plants used as food than in animals, because most of the larger amounts of residues in plant tissues are derived from foliar sprays applied sometime before harvesting. If the timing of such sprays is carefully controlled, residues in the plant tissues can be minimized. Nevertheless, it is very difficult to produce crops that contain no insecticide residues without serious losses due to pests. In the United States there are governmental tolerances set for the permissible amounts of residues for each crop and insecticide; these may vary from zero to several parts per million. Based on such tolerance limits, and experimental evidence on accumulation in human beings, FAO/WHO have agreed on recommended acceptable daily intakes of pesticides;[293] such calculations involve a very considerable safety factor (Table 20, see page 47).

During the last two years, surveys of the amounts of pesticides occurring in human diets have been made in the United States, Great Britain, Canada, and other countries. Some of the earlier investigations studied the amounts of pesticides in complete prepared meals,[222] but more recent ones have examined random samples of foods offered for sale to the public.[2] These are termed "total diet studies" and have been defined by FAO/WHO as "studies designed to show the pattern of pesticide residue intake by a person consuming a 'typical diet'." The data produced in these surveys are extensive, and could not be given here in full, so the three most recent studies, one from Great Britain and two from the U.S.A., have been considered, and the average residues found for each group of foodstuffs are listed in Table 21 (see page 48).

Duggan[69] studied residues in food in the United States between 1963 and 1967, examining 25,000 samples annually based on market-basket samples. Corneliussen[47] based his studies from 1967 to 1968 on samples collected from 30 markets in 27 cities, and discussed regional differences. In Great Britain, Abbott et al.[2] used 20 colleges which supplied samples of the food, for analysis, for their survey.

The most frequently detected residues in all the surveys were DDT, dieldrin, and BHC, and in all food groups the residues of DDT, and related compounds, tended to exceed the total of all the other residues which occurred. The foods which contained the greatest amounts of residues were the fats group, with the second largest residues in the meats group. These results are not surprising in view of the lipophilic nature of these chlorinated hydrocarbon insecticides which are concentrated into animal tissues. There appeared to be only small differences in the amounts of residues in food in the United States between the two surveys; if anything, there was a slight decrease in the amounts occurring in the more recent survey. There were considerable discrepancies between the residues which occurred in the United States and those from Great Britain. No traces of heptachlor, endrin, and toxaphene were found in British food, whereas they were quite common in the food samples in the United States, more being found in the more recent survey than in the earlier one. This can be related to differential usage of these insecticides in the two countries. In Great Britain, the smallest residues were in milk, but in the United States the second largest residues occurred in milk and dairy products; this is hard to explain.

Taken as a whole, the residues were considerably lower, both in the United States and in Great Britain, than the legal tolerances set by

TABLE 20

SUMMARY OF ACCEPTABLE DAILY INTAKES, RECOMMENDED TOLERANCES, TEMPORARY TOLERANCES & PRACTICAL RESIDUE LIMITS

Compound	Acceptable daily intake (max) mg/kg body weight	Recommended tolerance ppm*	Practical residue limits ppm†
Aldrin & dieldrin	0.0001	None	Milk 0.003 Meat 0.2 (on fat basis)
DDT	0.01	Berries (conc.) 1.2 Nuts (shelled) 1.0 Citrus 4.0 Other tree fruits 7.0 Vegetables 1.0 - 7.0 Meat, fish, poultry 7.0 (on fat basis)	Vegetables 0.05 Milk 0.005 Milk products 0.2 (on fat basis)
Gamma BHC	0.0125	Cereals 0.5 Vegetables 3.0 Small fruits 3.0 Milk products 0.1	Milk 0.004 Meat & poultry 0.7 (on fat basis)
Heptachlor & H. epoxide	0.0005	Root vegetables (other than potatoes) Head lettuce Spinach Leafy vegetables (from application to seed and soil only) 0.1	Meat 0.05 (on fat basis) Potatoes 0.05 Milk 0.002 Milk products 0.025 (on fat basis)

From WHO/FAO Tech. Report 370, 1967 (292)

* mg/kg
† mg/kg or mg/l

47

TABLE 21

INSECTICIDE RESIDUES IN HUMAN DIET
RESIDUE (pp 10^{-6})*

Location	Food	DDT & related	γ BHC	Aldrin	Dieldrin	Heptachlor & H. epoxide	Endrin	Toxaphene
U.S.A. Duggan 1968 (69)	Large fruit	0.012	t	0.003	t	-	t	-
	Small fruit	0.08	t	t	t	t	t	t
	Grain & cereal	0.005	0.008	t	0.002	-	-	0.005
	Leaf vegetables	0.025	t	t	t	t	t	t
	Fruit vegetables	0.048	0.002	t	0.002	t	t	-
	Root vegetables	0.008	-	-	0.001	-	t	-
	Peas & beans	0.01	t	t	t	t	-	-
	Eggs	0.02	t	-	t	t	-	-
	Milk	0.13	0.01	t	0.04	0.03	-	-
	Butter & cheese	0.22	0.15	t	0.01	t	-	-
U.S.A. Corneliussen 1969 (47)	Fruits	0.0094	0.005	0.0015	-	0.0002	-	0.047
	Garden fruits	0.0398	0.001	0.0005	0.003	0.0005	-	-
	Grain & cereal	0.005	0.0048	0.0007	0.0008	0.0002	-	-
	Leaf vegetables	0.0214	0.0014	0.002	0.003	-	-	0.011
	Fruit vegetables	0.0398	0.001	-	-	-	-	-
	Root vegetables	0.0002	0.001	-	0.002	0.0007	-	-
	Peas & beans	0.0204	0.001	0.002	0.0002	-	-	-
	Dairy products	0.112	0.008	-	0.0114	0.0114	-	-
	Oils & fats	0.0292	0.037	-	0.026	0.002	0.03	-
	Meat, fish & poultry	0.281	0.0172	-	0.0222	0.0102	-	0.062
Great Britain. Abbott et al. 1969 (1)	Fruits & preserves	0.025	0.005	-	0.0015	-	-	-
	Cereals	0.0175	0.0085	-	0.0025	-	-	-
	Green vegetables	0.0125	0.0055	-	0.0025	-	-	-
	Root vegetables	0.0055	0.0035	-	0.002	-	-	-
	Milk	0.0035	0.003	-	0.002	-	-	-
	Fats	0.2075	0.059	-	0.024	-	-	-
	Meats	0.0495	0.0165	-	0.009	-	-	-

* mg/kg
t = trace

48

both of these governments, or the ones recommended by FAO/WHO (Table 20). It seems that at the present time the amounts of insecticides being consumed by human beings are insufficient to be hazardous, but are certainly large enough to require continued vigilance in monitoring them, and need increased efforts to reduce them in coming years.

Pesticide Residues in Human Beings

Relatively soon after chlorinated hydrocarbon insecticides became commonly used, Howell[133] reported DDT residues in the fat of a human being and since then there have been many similar reports not only of DDT, but of most of the other chlorinated hydrocarbons in human body tissues.[17 27 122 223 268] Table 22 (see page 50) summarizes the amounts that have been reported from human tissues in surveys from many parts of the world; it is not exhaustive, but includes sufficient data to show that insecticide residues are very common in human bodies. Clearly the amounts of DDT which occur in human tissues greatly exceed those of all other chlorinated hydrocarbon insecticides, often reaching many parts per million in the tissues. Probably the average amounts of DDT present in human tissues at the present time lie between five and ten parts per million. The residues most commonly found, other than DDT, are aldrin and dieldrin. Residues of heptachlor occurred in the bodies of people from the United States, Canada, Holland, and France, but not in people from any other country. The largest residues of DDT and related compounds occurred in samples from bodies of human beings from the United States and India. No endrin was reported in any of the surveys.

Robinson[221] raised queries as to the validity of these data before considering their significance. He pointed out that accurate analytical methods are necessary, that knowledge of residues in different tissues is essential, that samples should be random and representative of the population, and finally, that particular groups in the population may be at greater risk than others. Most of these objections are now accounted for in that more sensitive and reliable analytical methods are being used, different body tissues are being analyzed, and the amounts of residues in people that are more ex-

posed to pesticides than the general population are being studied. There is still the bias that many analyses are made on tissues from bodies at autopsy, and often the cause of death may make such samples possibly unrepresentative of the normal population, but Robinson et al.[223] reported no significant differences between analyses made on specimens from biopsy and those from necropsy.

Hayes and Curley[120] studied concentrations of dieldrin in the tissues of men working in a factory that manufactures dieldrin and aldrin, and found that the average concentration of dieldrin in the workers, even if they worked only in the offices, was nine times greater than that in the general population. No endrin was found, although this insecticide was also manufactured in the factory. There was no meaningful correlation between the occurrence of dieldrin in samples, and average sick leave of the men per year. Correlations between hours of exposure to the insecticides, and the concentrations of residues in the tissues, were poor and concentrations did not seem to increase above certain levels, however long the exposure continued. There were correlations between the amounts of dieldrin in tissues and the amounts in urine, so urine samples or blood samples may be a convenient way of monitoring the amounts of pesticides in human tissues when more data are available. The concentrations of dieldrin were about 250 times more in samples of fat than in the blood of the men sampled.

In another investigation,[164] the amounts of DDT in the tissues of men with 11 to 19 years of exposure to DDT in a factory which produced the chemical exclusively were studied. The fat of these men contained 38 to 647 ppm of DDT, as compared with an average of about 8 ppm for the general population, and their daily intake was about 18 mg per man per day, as compared with about 0.04 mg for normal people; none of these men showed any demonstrable ill-effect

Hunter et al.[138 140] studied the pharmacodynamics of dieldrin in 13 human volunteers who took oral doses of dieldrin for prolonged periods. The doses, which were either 10, 50, or 211 micrograms of dieldrin daily, caused no signs of ill health in any of the subjects throughout the eighteen months of the experiment, nor were there clinical symptoms of any

TABLE 22

PESTICIDE RESIDUES IN HUMAN ADIPOSE TISSUES

RESIDUES (pp 10^{-6})*

Location	Reference	No. of specimens	p,p'-DDE Min.	p,p'-DDE Max.	p,p'-DDE Mean	DDT & related compound Mean	γ BHC Min.	γ BHC Max.	γ BHC Mean	Aldrin & Dieldrin Min.	Aldrin & Dieldrin Max.	Aldrin & Dieldrin Mean	Heptachlor & H. epoxide Min.	Heptachlor & H. epoxide Max.	Heptachlor & H. epoxide Mean
Great Britain	Egan 1965 (86)	66	0.1	4.8	2.0	3.1	t	1.0	0.34	0.02	1.08	0.21	-	t	t
"	Robinson et al. 1965 (223)	100	0.4	8.3	2.3	3.3	t	0.03	0.015	0.02	1.08	0.23	-	-	-
"	Cassidy et al. 1967 (38)	101	t	8.0	1.5	2.6	t	0.8	0.16	t	1.8	0.23	-	-	-
"	Robinson 1969 (221)	16	2.0	9.6	4.5	6.0	-	-	-	0.08	0.68	0.27	-	-	-
"	"	53	-	-	-	-	-	-	-	0.07	0.6	0.21	-	-	-
"	Hunter et al. 1963 (139)	131	-	-	-	-	-	-	-	0.02	1.2	0.21	-	-	-
Holland	Wit 1964 (289)	20	0.08	19.0	5.5	7.1	0.05	0.21	0.1	0.05	0.50	0.17	0.004	0.03	0.009
"	de Vlieger et al. 1968 (268)	11	-	3.3	1.7	2.0	-	-	-	-	-	-	-	-	-
France	Hayes et al. 1963 (121)	10	-	11.1	3.2	5.2	-	0.15	0.06	0.03	0.99	0.45	-	1.37	0.21
Italy	del Vecchio & Leoni 1967 (267)	22	2.8	27.6	7.5	8.2	-	-	-	-	-	-	-	-	-
Israel	Wasserman et al. 1967 (276)	71	0.3	32.9	5.6	4.6	-	-	-	-	-	-	-	-	-
"	"	133	1.0	40.6	9.9	8.2	-	-	-	-	-	-	-	-	-
India	Dale et al. 1965 (54)	24	2.0	23.0	11.6	27.8	0.2	4.9	1.7	0.01	0.36	0.03	-	-	-
"	"	11	0.4	23.9	6.4	11.9	0.1	3.2	0.9	-	0.19	0.06	-	-	-
New Zealand	Brewerton & McGrath 1967 (23)	52	0.63	11.0	3.8	5.3	-	t	t	0.01	0.77	0.27	-	-	-
Australia	Bick 1967 (17)	53	0.25	3.5	0.93	1.7	-	-	-	0.03	0.43	0.05	0.03	1.45	0.24
U.S.A.	Hayes et al. 1965 (122)	25	0.6	21.2	6.85	10.0	0.03	2.43	0.06	0.03	1.15	0.29	-	-	-
"	Dale & Quinby 1965 (53)	28	0.5	10.4	3.8	4.9	-	0.7	0.2	-	0.36	0.15	0.03	0.61	0.1
"	Zavon et al. 1965 (297)	64	0.9	14.7	4.6	7.0	-	-	-	0.07	2.82	0.31	-	-	-
"	Fiserova-Bergerova et al. 1967 (93)	42	1.5	19.0	6.7	9.8	-	-	-	-	0.7	0.22	-	-	-
"	Hoffman et al. 1964 (129)	994	t	56.1	7.0	7.6	-	12.3	0.48	-	1.39	0.14	-	0.74	0.16
"	Edmundson et al. 1968 (74)	146	3.6	12.5	7.0	10.3	-	-	-	0.05	0.77	0.22	-	-	-
Canada	Brown 1967 (27)	47	0.6	6.8	2.66	4.0	-	0.18	0.06	-	0.53	0.16	-	0.4	0.07

* mg/kg

abnormalities. After 18 months, the body burdens of those individuals receiving 50 micrograms of dieldrin daily were four times greater than those of the general population, and those treated with 211 micrograms were ten times those of the general population. The most interesting feature was the non-linear relationship between the concentration of dieldrin in the blood and the time of exposure to the insecticide; amounts in the blood increased during the first ten months of exposure, and, thereafter, increases were very small. It was concluded that there is a finite upper limit to the storage of dieldrin by each person, being characteristic of that person, and his particular daily intake only. There were significant correlations between the amounts of dieldrin in the blood, and adipose tissues.

Thus the available evidence shows that, although amounts of chlorinated hydrocarbons in human tissues are widespread, they do not continue to increase linearly with amounts in the environment, but reach an equilibrium value. This plateau level might perhaps increase if the amounts of chlorinated hydrocarbon insecticide residues in the environment continue to increase.

Hunter[137] calculated what he considered to be the maximum allowable concentrations of chlorinated hydrocarbon insecticides in human tissues. There is no evidence that present levels of chlorinated hydrocarbon insecticide residues in human tissues have caused any ill-effects to people that carry them in their tissues, although it may be too early to assess the long-term effects. It has been shown[119] that these chemicals can be transferred from the mother to the fetus of mammals including man, so that babies may be born with some chlorinated hydrocarbon in-

secticides in their tissues.

The relative importance of the possible sources of persistent insecticide residues in human tissues has not been fully evaluated. It seems likely that the food is the most important source, particularly meat and dairy products, which usually contain the greatest amounts of insecticide. Campbell et al.[35] estimated that approximately 90% of the uptake of DDT into human beings could be accounted for by food intake, but Kraybill[160] considered this a little high and stated that 85% comes from food, the remainder coming from air, water, aerosols, cosmetics, and clothing. Brown[27] considered that chlorinated hydrocarbon insecticide residues in potable water supplies were an important source of human contamination, but the amounts reported from drinking water hardly bear this out. Hayes[119] estimated that the average amounts in drinking water taken in per day would be 0.046 micrograms per person, with a possible maximum of 2.0 micrograms. In some areas, the amounts respired in the air may contribute appreciably. Table 4 compares amounts of chlorinated hydrocarbon insecticides taken in with air, with those taken in with food, for one area of the United States, but this is probably higher than normal, and more realistic calculations are those of Tabor[248] (2.0 to 8.0 micrograms per day).

Human beings seem to be adapting to the amounts of pesticides currently in the environment. Robinson and Roberts[226] considered that there was some evidence of a decline in the concentrations of dieldrin in man and the environment in Great Britain since 1965; if this continues, and is accompanied with a similar decline in DDT residues, it appears that man will not be exposed to very serious hazards.

THE POSSIBLE CONTROL OF PERSISTENT PESTICIDES IN THE ENVIRONMENT

The Use of Alternative Pesticides

The chlorinated hydrocarbon insecticides have been used for less than 30 years, but the quantities used have been colossal; in the United States alone, more than a million tons of DDT and 600,000 tons of aldrin and dieldrin have been manufactured. Unless alternative methods

of pest control are found, and existing pesticides still used at the present rates, further accumulations in the physical and biological environments will continue. Already, however, many insecticides have been developed which are much less persistent, and will kill many of the pests against which the chlorinated hydrocarbon insecticides have been used. There has been an added impetus to the search for such chemicals in Great Britain since 1964, when restrictions were placed on the use of aldrin and dieldrin.[190] The expected crop losses, if such restrictions were made, were estimated at the time.[246] Although data are not yet available to show how accurate these predictions were, it seems probable that losses were less than anticipated, probably because new insecticides were rapidly developed, being mainly organophosphates or carbamates, and all with a persistence of less than one year. These new chemicals are not only less persistent in the environment, but also do not accumulate in animal and plant tissues as do chlorinated hydrocarbons. Unfortunately, many of them have higher mammalian toxicities than the chlorinated hydrocarbons, and are rather more of a hazard to the pesticide operator.

In the United States, there have been no more than local bans on the use of chlorinated hydrocarbon insecticides, but the strict application of food tolerances has necessitated a search for less persistent insecticides as substitutes. If the U.S.D.A. recommendations for insect control are compared for 1969 and 1959, it is striking how many new chemicals are now recommended in place of the chlorinated hydrocarbons. The main problem of finding alternatives for persistent insecticides is not that suitable insecticides cannot be developed but is, rather, economic because the persistent insecticides can be made and sold so cheaply, and one dose gives protection against soil pests for several seasons. In the United States there is also the problem of controlling insects that are household or urban pests, because these cause much more trouble than in Europe, and insecticides toxic to man cannot be used to control such pests. There is considerable pressure in the United States to ban the use of DDT completely; now this is proposed, residues in the environment will no doubt fall, but alternative control measures are essential for some pests.

Better Use of Pesticides

Many of the problems that have arisen through the use of persistent pesticides are due to careless use of these chemicals. Too often, particularly in the United States, large areas of land have been indiscriminately sprayed to control forest insects, mosquitoes, or agricultural pests. Such sprays fall on all parts of an ecosystem, and it is hardly surprising that residues occur throughout the environment after such treatments. Such spraying operations should be severely limited, even with non-persistent pesticides. Nevertheless, the efficiency of spraying insecticides has increased greatly in recent years, and less and less water is required, together with a reduction in the amount of chemical applied. It is now possible, using sophisticated spraying equipment, to apply ½ lb of active ingredient of insecticide per acre undiluted with water.

For many years, insecticides were broadcast over the surface of the soil every year, then thoroughly incorporated into the soil with a plough or rotovator. Such treatments were very uneconomical, and used much more insecticide than was really necessary to control a particular pest. If treatment of the whole soil is essential, then it is much better to calculate a "topping-up" dose which will bring existing residues back to a sufficient level to control pests. Decker et al.[59] described how the amounts of aldrin applied to control the root pests of corn could be calculated in this way. It is much better, however, to use localized treatments which place the insecticide exactly where it is required; for example, by using a seed-dressing of insecticide, which applies a very small quantity of insecticide close to the seed. In recent years it has been found that insecticides have been applied inefficiently to seeds in the past, but it is hoped that current research will produce greatly improved "stickers," or other methods of ensuring a uniform cover of insecticide over the seed, and increase the efficiency and use of seed-dressings. Machinery has been developed in recent years which allows small doses of insecticides to be applied close to the seeds as they are sown; other machinery applies narrow bands of insecticide in the rows of the crop. Some pests can be controlled by dipping the roots of transplanted crops in insecticides before planting out.

The efficiency of insecticides which have only a short life in soil, can be extended to cover a whole growing season by applying them adsorbed on the surface of inert granules which gradually release the insecticide; different granules release insecticides at different rates, so that most suitable ones can be chosen for a particular pest. There is also a new technique of enclosing particles of non-persistent insecticides in micro-capsules, which gradually break down to release the insecticide into the soil.

Finally, the possibility of using insecticides in eradication programs should be considered. These may well be expensive but the ultimate benefits of freedom from a pest may be great. Such programs usually succeed only against pests introduced to countries other than that of their origin, or in delimited geographical areas.

Removal of Pesticides

Once chlorinated hydrocarbon insecticides become thoroughly incorporated into soil, they are adsorbed on to soil fractions and persist much longer than if they merely lie on the soil surface. It has been estimated that these chemicals disappear at least ten times faster if left exposed to weathering and evaporation, than if cultivated into soil.[76] Cultivations of soil around crop plants that have been sprayed with persistent insecticides should therefore be kept to a minimum; the use of herbicides to control weeds makes this a practical proposition. The tendency, in modern agriculture, to use minimal tillage techniques, does keep cultivation to a minimum; if the completely non-tillage techniques, that have shown promise in Great Britain, are generally accepted, even less insecticide will become mixed into the soil. Lichtenstein et al.[173] investigated the possibility of accelerating disappearance of insecticides from soil by frequent cultivations which bring the residues to the surface, but although this did increase the loss of the insecticides, it certainly would not be an economic proposition to cultivate as often as they did—every three days.

Lichtenstein et al.[176] investigated the possibility of removing residues from active sites in soil by adsorbing them on activated carbon. The addition of carbon, at a rate of 1,000 ppm to a quartz sand, reduced the penetration of aldrin into pea roots by 96%. To achieve a similar reduction in a loam soil, 4,000 ppm were necessary, and to stop dieldrin or heptachlor entering pea roots, 2,000 ppm were required. Obviously, adding carbon might be a practical proposition in very badly insecticide-contaminated fields, but it is doubtful if the cost would

be justified. Furthermore, the method does not remove the insecticide, but merely inactivates it, and the presence of the carbon in the field would also inactivate later applications of other insecticides and herbicides.

It is becoming increasingly clear that microorganisms are very important in breaking down persistent insecticides in soil, and that particular microorganisms cause much more insecticide degradation than others. It is an intriguing possibility that, by the addition of cultures of suitable microorganisms or of substrates to soil, the disappearance of persistent pesticides could be expedited.

There is good evidence that insecticides break down more rapidly in soils that have previously contained residues of that insecticide, possibly because the appropriate microorganisms are present in larger numbers in such soils.

Guenzi and Beard[107] showed that DDT broke down to DDD much more rapidly under anaerobic conditions, and even faster if 1000 ppm ground alfalfa foliage were added to soil. They speculated on the possibility of flooding agricultural land as a means of keeping the soil anaerobic and speeding up DDT breakdown.

There is little evidence that chlorinated hydrocarbon pesticides reach drinking water in other than minute trace amounts, but it is important that, if drinking water should become contaminated, the pesticides be removed. Robeck et al.[217] tested the effectiveness of various filtering methods in removal of dieldrin, endrin, lindane, and DDT, and they found that conventional treatment with coagulation, followed by sand filtration, varied in effectiveness. Almost all of the 10 ppm of DDT were removed by this treatment, but less than 20% of the lindane and parathion disappeared. Softening with lime and soda ash, with an iron salt as coagulant, did not improve on the results obtained with alum coagulation alone; in fact, less was removed. Treatment with chlorine or potassium permanganate had no effect on the insecticides either. At large and impractical concentrations, ozone decreased amounts of chlorinated hydrocarbon insecticides in the water, but the by-products were not identified. The addition of activated carbon to the filtration system greatly decreased the amounts of insecticides in the water, but the amounts of carbon required were so large as to make the operation of no more than academic interest. Nicholson et al.[206] found that carbon adsorption removed DDT, but not BHC or toxaphene, although Cohen et al.[42] did obtain some reduction of toxaphene in water, with 1 to 9 $pp10^{-6}$ of activated carbon. Robeck et a.[217] also found that lindane could be removed by activated charcoal; about 29$pp10^{-6}$were required to decrease a lindane concentration of 10 $pp10^{-12}$ by 90%, but 9 $pp10^{-6}$ decreased the concentration of lindane from 1 to 0.05 $pp10^{-12}$. They did not find that oxidants such as ozone or potassium permanganate were effective.

Thus, although current water treatment measures do not greatly decrease the amounts of chlorinated hydrocarbon insecticides in water, there seem to be possibilities of increasing the efficiency of extraction of pesticides from the water.

It now seems clear that runoff from agricultural land is an important source of pesticides in rivers and streams; probably the insecticides are carried in the water on particles of soil. Controlling such runoff would be one way of decreasing contamination of water. On flat agricultural areas there is not much of a problem, but in watersheds, there is often a great deal of surface water movement in times of heavy rain. Draining such areas might be one way of reducing surface runoff. Insecticides should not be applied on sloping land that is close to water, and when insecticides are applied, there should be an untreated buffer strip between the treated area and the water.

It is important to minimize the amounts of pesticides in food for human beings. A considerable proportion of the pesticides in food of plant origin can be removed in processing, e.g., by washing, peeling, or heating.[244] Washing, which is probably the most important, can be improved by using warm water and adding suitable detergents to the washing water. Pesticides can be removed from vegetable oils by steam stripping, under low pressure, to induce forced volatilization of the residues. The concentrations of pesticide residues in milk can be lessened by various drying processes, but the efficiency of these processes differs between different insecticides.

Lessening the concentrations of insecticide residues in livestock is more difficult, but has been achieved for dieldrin by feeding cattle with charcoal[244] and the procedure warrants further study. Certain drugs, e.g., barbiturates, have also increased the excretion of DDT from cattle. These methods are, however, poor substitutes for minimizing exposure to the insecticides.

Substitutes for Pesticides

There have been many new approaches to pest control during the past decade. Firstly, the concept of integrated control has been developed; this implies that when control measures are used they should integrate cultural and ecological control measures with insecticidal ones, to obtain maximum effect with minimum use of insecticides. In the past, too often, insecticides have been applied to crops as blanket treatments, using very empirical doses without serious consideration of the ecology of the pest; often they have been used merely as insurance.

Some pest control methods do not require the use of pesticides; for example, such methods as timing of sowing dates, timing harvesting, rotating crops, using additional cultivations to kill soil pests, and the use of resistant varieties of crops have long been known, and were widely practiced before the advent of the chlorinated hydrocarbon insecticides.

Biological control of pests has been an attractive method of pest control for many years, but there are several limitations on achieving the classical form of biological control of pests, which consisted of introducing an inimical biological agent into the territory of the pest. Usually, biological control has been successful only with pests introduced from other countries, and in areas isolated topographically or geographically. In recent years there have been extensive programs investigating the possibilities of introducing pests, predators, or pathogens to areas with serious pest problems. Some have succeeded but, in general, such methods cannot yet be a practical substitute for insecticides.

There are now extensive research programs into the development of insect attractants and repellants.[144] Insects are very sensitive to odors and very small quantities of attractants can be used to lure them to traps. The attractants, which are usually food or sex-based, may be effective at distances of up to a mile and are usually highly specific, attracting only a few closely related species, and then often only males. Attractants can be used either as an index of the presence of a pest or to trap the pests and thus remove them. Attractants were successful in helping to eradicate the Mediterranean fruit fly from Florida. The possibility of using chemosterilants on insects that are trapped by attractants, then releasing the sterilized insects, is also being investigated for certain pests.

Chemosterilants and the use of radiation to sterilize insects are also being extensively investigated. These methods depend on the ability of insects to mate once only; if this mating is with another insect that has been sterilized by chemicals or radiation, there will be no offspring. The classical success of this technique has been with the screw-worm fly which was previously a serious pest of cattle in the U.S.A.,[156] two other insects, the melon fly on the island of Rota in the Pacific, and the Oriental fruit fly in Guam have been successfully eradicated using this method. Research into the use of this method on pink bollworm and tsetse fly is now in progress, but the technique has serious limitations, and is of application to only a small minority of pests and areas.

In recent years there has been most encouraging work on the development of biological pesticides which depend on the dissemination of spores of pathogenic microorganisms; this is not new, but greatly increased research is beginning to yield numerous pathogens that may be suitable for this kind of control. Such methods are very attractive in being confined to the application of biological agents which are unlikely to cause hazards because they persist in the environment. There is the slight hazard that some of these pathogens might be able to attack more than one pest or other animal, but this is unlikely. The sorts of pathogens usually used are fungi and bacterial spores, but there are indications that it may also be possible to disseminate viruses to control pests.

Governmental Control of the Use of Pesticides
The use of many chemicals is currently restricted by law, and insecticides are no exception to this rule. There exists already in the United States, Great Britain, and other countries, governmental machinery for controlling the marketing of pesticides. In the United States there are stringent regulations and requirements on pesticide manufacturers before they are allowed to market any pesticide. There are also carefully worked out tolerance limits on the amounts of pesticides that contaminate the environment.

So far, however, outright bans on insecticides are rare, although voluntary restrictions were placed on aldrin and dieldrin in Great Britain in 1964. Many countries are now, however, considering banning DDT, aldrin, and dieldrin, and recently both Canada and Denmark have banned DDT. The United States Department of Health, Education, and Welfare has recommended that DDT should be banned. It was very difficult to make this decision because the good resulting from the use of DDT, particularly in controlling tropical diseases, has been great, and it may well be that restricting its use to controlling certain pests would be a better compromise.

There is certainly scope in promoting the education of insecticide users and controlling the labeling and instructions on insecticide containers. Many of the persistent insecticide residue problems we now have are due to misuse of insecticides. Restricting application methods for chlorinated hydrocarbon insecticides, e.g., banning aerial spraying, might also be valuable.

To summarize, it does not seem that the present situation concerning pesticides in the environment is too serious. Many resources are being expended to develop alternative chemicals or methods of pest control, and the environment is already being monitored for residues. Investigations into the hazards to animal and human populations by insecticide residues in the various parts of the environments are proceeding. Many governments are taking action to restrict the use of chlorinated hydrocarbon insecticides. It seems, however, that the monitoring program could well be improved and made uniform and extensive.

APPENDIX

Legislation on Persistent Pesticides during 1969/70

The ecological hazards associated with the use of persistent pesticides have caused international concern during 1969 and 1970, and many countries formed committees to study these problems or introduced legislation during this period. Such actions will greatly affect the usage of these chemicals and, hence, the extent the environment will become contaminated in the future; so it is important to consider them in some detail by countries or geographical areas.

United States

During 1969 various states were considering restricting the use of persistent pesticides. A ban on the use of DDT for a period of one year was already in force in Arizona, and Wisconsin was considering taking similar action. Bills of the same kind were being considered in several other states, including Washington, Maine, Connecticut and Montana. California, which has been a pioneer in developing regulatory plans for pesticide usage, was also considering legislation, particularly in respect to DDT. In Illinois, a bill which proposed banning the use of DDT was passed in the House but later rejected by the Senate Agricultural Committee.

The Michigan State Commission on Agriculture stopped issuing new licenses and canceled existing licenses for the sale of DDT, in April 1969. This occurred after the Food and Drug Administration seized 28,000 lb of Coho salmon, which contained residues of up to 19 ppm of DDT and its breakdown products. In June 1969, the Agricultural Department of the state of Michigan stated that all uses of DDT were to cease forthwith.

The U.S. Department of Agriculture issued a directive in April, which stated that approval for the use of DDT on cabbage and lettuce for certain purposes had been withdrawn, in order to prevent residues of this insecticide occurring in these crops, although DDT could still be used in the early growing stages.

A federal commission, which was asked to gather and evaluate all available evidence on both the benefits and risks of using pesticides and to pay particular attention to DDT, was appointed by the U.S. Department of Health, Education, and Welfare in April 1969 and asked to report back to the Department within six months. The initiation of this investigation was partially motivated by reports that Americans had an average body burden of 12 ppm of DDT.

A committee set up by the U.S. National Research Council reported to the U.S. Department of Agriculture in May 1969.[317] They concluded that persistent chemicals should not be needlessly released into the atmosphere but were essential in certain situations. They could foresee no decrease in the use of pesticides including the organochlorines. Other conclusions were that available evidence did not indicate that present levels of pesticide residues in man's food and environment produced an adverse effect on health, but that persistent pesticide residues do have an adverse effect on some species of wild animals, and threaten the existence of others.

The committee recommended that action be taken at international, national, and local levels to minimize environmental contamination, that further research on long-term ecological effects on man and other mammals be intensified, that public funds for research on chemical methods of pest control be increased, and that the National Pesticide Monitoring Program should be reviewed.

The U.S. Secretary of the Interior introduced a bill during May 1969 which would give him powers to restrict or prohibit the use of pesticides in water if they presented a hazard.

A bill was introduced to the California legislature in June which would give the Director of Agriculture authority to refuse to register, or to cancel the existing registration of, any economic poison that had demonstrable side-effects. Although this bill was rejected in August, the California Department of Agriculture introduced regulations which restricted the use of DDT or DDD in households and gardens, or in dust form for use in agriculture, with the aim of gradually phasing out the use of DDT altogether. Finally, as from January 1, 1970, the California legislature completely banned the use of DDT in homes and gardens, for crop dusting, and for treating farm livestock. In August, the New York State Pesticide Control Board proposed that DDT and certain other persistent pesticides should be sold only to registered applicators and people holding permits. It suggested that legislation should be introduced which would eliminate the use of DDT completely by 1971 except in emergencies.

The U.S. Department of Agriculture suspended part of its federal pest control program in July 1969, pending a review within 30 days. The chemicals which they stopped using included DDT, dieldrin, endrin, aldrin, chlordane, lindane, heptachlor and benzene hexachloride. The U.S. Senate added to the Department of Agriculture's appropriations an amendment which prohibited the use of particular pesticides by the federal government, in states where those pesticides were banned by law or regulation, or even their use in neighboring states, if this might substantially affect a state where such legislation existed. This review was completed in the middle of August, and the U.S. Department of Agriculture decided that for federal-state programs of pest control, some of the persistent pesticides would be replaced by less hazardous compounds. Examples that they specifically announced were that chlordane would be used for control of Japanese beetles (*Popillia japonica*) and European chafers (*Amphimallon majalis*), instead of dieldrin or heptachlor, and, where possible, chlordane would be used to control white-fringed beetles (*Graphognathus leucoloma*) instead of dieldrin. The use of dieldrin was to be restricted, although it could still be used under strict control when it was required to protect shipments of nursery stocks between states. Two federal bills were also introduced in August, one for a four-year suspension of the use of DDT, aldrin, dieldrin and endrin; the other proposed that DDT should be completely banned from June 30, 1970.

The subcommittees of the federal commission, appointed by the U.S. Department of Health, Education, and Welfare, in April 1969, produced reports in November 1969,[318] which included the following observations:

Uses and Benefits Subcommittee

1) The production and use of pesticides in the United States is expected to grow at an annual rate of approximately 15%, and insecticides will more than double in use by 1975.

2) Although the use of DDT is declining in the United States, an increasing quantity is being purchased by AID and UNICEF for foreign malaria programs.

3) Most persistent pesticides other than DDT have continued to decline in use since 1957, a trend that should continue. There will continue to be a need for use of persistent materials for the control of selected pest problems.

Contamination Subcommittee

1) Much contamination and damage result from the indiscriminate, uncontrolled, unmonitored and excessive use of pesticides. The careful application and use of techniques currently available can be expected to reduce contamination of the environment to a small fraction of the current level, without reducing effective control of the target organisms.

2) A single agency should initiate effective monitoring of the total environment on a continuous system.

3) Aerial spraying should be confined to conditions that preclude drift.

4) The use of low-volume concentrated sprays should be encouraged.

5) Alternatives to persistent pesticides should be developed.

Effects on Non-Target Organisms other than Man Subcommittee

1) Persistent chlorinated hydrocarbon insecticides are causing serious damage to certain birds, fish and other non-target species, and they should be progressively removed from general use over the next two years.

Effects of Pesticides on Man Subcommittee

1) The consequences of prolonged exposures of human beings to persistent pesticides cannot be fully elucidated at present, the only unequivocal consequence being the acquisition of residues in tissues and body fluids. No reliable study has revealed a causal association between the presence of these residues and human disease.

2) The uneven distribution of residues among the populations of different geographical areas casts serious doubt on the accepted beliefs that food is the predominant source of DDT residues and that the entire general population has reached equilibrium as regards acquisition of such residues.

The Commission made 14 recommendations to the Department of Health, Education, and Welfare. The more important of these included:

1) The uses of DDT and DDD should be restricted within two years to those which are essential for the preservation of human health and welfare, and approved unanimously by the Secretaries of the Departments of Health, Education, and Welfare, Agriculture, and the Interior.

2) The uses of certain other persistent pesticides should be restricted to specific essential uses which involve no hazard to human health or to the quality of the environment.

3) Suitable standards should be developed for permissible pesticide content in food, water and air, and other aspects of environmental quality.

They also recommended the following measures. There should be closer cooperation between various federal departments. A pesticide advisory committee should be created to evaluate information on the hazards of pesticides to human health and environmental quality. A clearing house should be established for pesticide information. Federal support for research on all aspects of pest control should be increased. Incentives should be provided to encourage industry in the development of more specific pest control chemicals. The legislation and regulation concerning labeling, definition of toxicity and warning of hazards should be reviewed. Regulations for the collection and disposal of unused pesticides, used containers and other pesticide contaminated materials should be developed. Participation in international cooperative efforts to promote safe and effective usage of pesticides should be increased.

The report summarized the results of research on the hazards of persistent pesticides, but the ways in which the various sub-committees dealt with their data differed consider-

ably. The summaries of effects of these pesticides on invertebrates, fish, birds and mammals were extremely short and considered only a small part of the extensive literature and research on the subject. The greatest part of the report (p. 229–677) was concerned with the effects of pesticides on man, with particular attention to carcinogenicity, mutagenicity and teratogenicity caused by these chemicals, and the literature on this subject was dealt with much more fully than that of other environmental hazards due to pesticides.

The U.S. Department of Health, Education, and Welfare supported the recommendations of the Committee, and as a result, on November 20, 1969, the U.S. Secretary of Agriculture ordered that all uses of DDT in residential areas should cease within 30 days, and he stated that similar action would be taken in respect to other persistent pesticides in 1970. No final decision on the use of DDT in agriculture was made, and it has been reported that the Secretary of Agriculture wants to retain DDT for use on crops where there is no adequate alternative. Final decisions on the scale of use of persistent pesticides remain to be made.

Canada

In November 1969, the Canadian government announced extensive restrictions on the use of DDT. This insecticide would not be registered for use to control insects in forests, parks and outdoor areas except under emergency conditions from January 1970. It was also withdrawn from use on 50 food crops, and was in future permitted to be used only on apples, blackcurrents, celery, mustard, radishes, rapeseed, raspberries, redbeet, strawberries, sugarbeet, turnips and tobacco. The accepted tolerance levels of DDT in food were also reduced.

Great Britain

The Advisory Committee on Pesticides and Other Toxic Chemicals, which reported back to the British Department of Education and Science, produced a document entitled, Further Review of Certain Persistent Organochlorine Pesticides Used in Great Britain,[304] in February 1970. In the previous Review[190] in 1964, the Committee had examined the risks arising from the use in agriculture (including gardening) and food storage, or aldrin, benzene hexachloride, chlordane, DDT, dieldrin, endrin, endosulfan, heptachlor, 'Rhothane' and toxaphene. They concluded that, at that time, there was insufficient evidence for a complete ban on any of these chemicals and proposed that their use should be allowed to continue, with certain restrictions on the use of aldrin and dieldrin. They also proposed that the use of all these insecticides, except benzene hexachloride, should be reviewed at the end of three years, with a possible view to their discontinuance.

In the 1970 Review, the Committee reached several important conclusions as to the effects of persistent insecticides on wildlife in Great Britain. Some of these conclusions were:

1) That they could not find evidence that the chemicals under review had caused widespread deaths among birds (except as seed dressings), although many wild birds contain small residues of these chemicals and their metabolites. There is the possibility that residues may cause stress in wild birds in the field, and there is experimental evidence that DDT can delay the attainment of sexual maturity in some species of birds and, hence, influence breeding success.

2) Although there was close correspondence between the use of DDT and the decrease in thickness of eggshells of peregrine falcons and sparrow-hawks, the evidence was not precise enough to establish a causal relationship with certainty.

3) Larger residues of dieldrin were found in predatory birds than in other species, and the declines in numbers of peregrine falcons and sparrow-hawks roughly correspond with the periods of usage of aldrin and dieldrin. Following the restrictions on the use of these chemicals, some population increases have been observed in the peregrine falcon, sparrow-hawk, and kestrel, and it seems likely that the golden eagle benefited from the ban on the use of dieldrin in sheep dips. DDT does not appear to have been a principal cause of decline in numbers.

4) Although residues of DDT and dieldrin and their metabolites have been found in wild mammals, there is no evidence that these have substantially affected the populations of these mammals. There is some experimental evidence that non-fatal doses of dieldrin can produce effects, the ecological significance of which has not yet been established, but these have occurred only when the dose was near to the lethal amount.

5) Little information is available from Great Britain on the effects of field applications of DDT on fish. Some episodes have come to light where mortality among fish due to dieldrin was attributable to industrial effluents. Residues of DDT in natural waters that have been examined, including those in heavily treated areas, are generally well below the threshold concentrations that would threaten survival or produce adverse effects on fish.

6) There is insufficient evidence to state whether or not organochlorine pesticides have affected marine invertebrates and fish. Nevertheless, the levels of insecticides in marine life in coastal waters are causing some concern. Limited studies in Great Britain suggest that pesticides reach the sea mainly by rivers.

7) DDT has been implicated only on rare occasions in bee-poisoning incidents.

8) DDT can have indirect ecological effects on arthropod populations; thus, pest mites in orchards benefit if the more sensitive parasites and predators are killed by this insecticide.

9) Food reaching the consumer contains consistently low levels of organochlorine pesticides. The daily intakes in microgram/kilogram body weight are, with the exception of dieldrin, very much lower than those proposed as acceptable by FAO/WHO Expert Committee on Pesticide Residues.

10) There is no evidence of harmful effects on man.

The committee made specific recommendations on the future use of persistent pesticides. In the 1964 Review they banned the use of aldrin, dieldrin and heptachlor in agriculture, horticulture, home gardens and food storage, except for certain uses. These included seed dressings on sugar beet and winter sown wheat, aldrin and dieldrin against cabbage root fly, narcissus bulb fly and wireworm in potatoes. In the 1970 Review these exceptions were withdrawn so that the use of aldrin, dieldrin and heptachlor was virtually banned for all purposes, and so was the sale of small retail packs of dieldrin. In addition, the use of endrin on apple trees, which was previously unrestricted, was banned.

The 1970 committee recommended that the use of DDT on grassland, brassica seed crops, peas (except for weevil control), raspberries, gooseberries, post-blossom for top fruit, and for aphid control on all crops, should cease. DDT should not be used in commercial dry cleaning, or for home-moth-proofing of woolen goods. Thermal vaporizers discharging DDT (and/or gamma benzene hexachloride) should not be used. DDT should not be used in home gardens or for control of insect pests in the home. Paints and lacquers containing DDT should not be used. Smoke generators containing DDT should be restricted to use in empty stowages, or stowages where any food present is protected by sheeting or impermeable packaging. Thermal vaporizers discharging DDT alone or in mixtures should not be used in rooms where food is sold, prepared or eaten, or when persons, animals, or unprotected foods are present.

They recommended that TDE ('Rhothane') should not be used post-blossom on apples and pears, on loganberries or raspberries. Gamma-benzene hexachloride should not be thermally vaporized in the living, eating or sleeping quarters or in buildings in which anyone is exposed to the continuously generated vapor for more than a normal eight-hour working period.

To prevent further hazards, the Committee recommended that residue levels in people, their diets and the environment should be kept under surveillance. The necessity for industrial uses of aldrin and dieldrin should be examined. Further research should be done on the toxicological and ecological significance to wildlife of pesticide residues in the environment. More work should be done on the factors associated with the field use of pesticides, and on cost/benefit analysis of the use of pesticides as an insurance against the possibility of crop loss.

Sweden

The National Poisons and Pesticides Board in Sweden announced that from January 1,

1970, the use of aldrin and dieldrin would be banned completely, and the use of DDT and lindane for domestic purposes in the home and garden were also banned. The other uses of DDT, in particular in agriculture, would be suspended for two years from the same date although, in exceptional circumstances, an exemption might be granted, e.g., for control of rapeseed beetles. During the period of the ban, research aimed at determining the effect of the restrictions would be initiated.

Denmark

In June 1969, the Poisons Board of Denmark stated that as from November 1, 1969, pesticides containing DDT would no longer be approved for general use, although dealers could dispose of existing stocks. Exceptions to the restrictions were to control moths, to treat woodwork and for some indoor uses. They also stated that they were considering amending the regulations concerning the use of lindane and would announce measures in the near future.

Norway

The Norwegian government announced that as from October 1970 all uses of DDT in horticulture, agriculture and forestry would be prohibited, and that they were considering a further ban on the use of DDT in households and to treat domestic animals, beginning from the same date. They also imposed a ban on the importation of DDT from abroad after October 1, 1969, in order to allow stocks to run down.

Italy

In November 1969, the Italian government announced that they would ban the use of DDT in agriculture for use on grains, legumes, fodder plants, vegetables, root crops, stored foods and certain fruits such as peaches, apricots, cherries and plums. They also proposed that regulations would be issued defining the period which must elapse between treatment and harvest of citrus fruit, olives and grapes. DDT was also to be banned for domestic use, treatment of stables and animals, and large-scale treatments against disease vectors. Limits were set on the permissible amounts of DDT in imported products, such as butter and cereals for human consumption, and in cattle food.

Comments on Other Countries

It is clear that the more advanced North American and European countries are taking action to restrict the indiscriminate use of persistent pesticides which may pollute the environment. The largest amounts of DDT, which appears to be the most hazardous of the persistent chemicals, are used in Africa and Asia, often in the underdeveloped countries, usually to control disease vectors. There is evidence of global transport of these pesticides and it seems extremely important that international cooperation in the restriction of the use of persistent pesticides should be initiated.

Persistent Pesticides other than Chlorinated Hydrocarbon Insecticides

The data and discussion in this review have been mainly confined to the organochlorine insecticides; this is because these chemicals are much more persistent and toxic than other pesticides, and also because their affinity for biological tissues has meant that they have been taken up into the bodies of animals and and plants, sometimes in large quantities.

Herbicides

These chemicals are much less persistent than the chlorinated hydrocarbon insecticides; most of them have a much lower mammalian toxicity and they do not usually concentrate in biological material. Most herbicides break down rapidly in soil, although some, such as atrazine, dichlorbenil, diuron, fenac, monuron, neburon, picloram, propazine, propham, simazine, trichloroacetic acid and 2,3,6-trichlorobenzoic acid may persist for a year or longer in soil.[150 315 322] The most persistent herbicides, in order of decreasing persistence, are propazine, picloram, simazine and 2,3,6-trichlorobenzoic acid. The times herbicides persist in different soils are influenced by the same factors as insecticides. Bailey and White[9] reviewed the soil characteristics which affect the persistence of herbicides in soil and showed that organic matter and clay content were very important. Unlike other pesticides, the presence of residues of herbicides in soil is usually immediately obvious, because of their phytotoxicity to weed and crop plants.

Diuron, and the closely related trifluralin, have been used in large quantities on cotton in the southern United States; but although there has been injury to cereals planted in fall, after treatment of the cotton crop, very rarely does diuron persist into a second growing season. However, there have been occasional reports of soybeans being damaged in the spring after treatment.[303] When atrazine has been used on corn, it has sometimes persisted into the following season, but its rate of breakdown is such that annual applications at recommended rates do not lead to accumulations. Large doses of some herbicides can persist in soil for several years, however; an area of soil treated with a very large dose of simazine at Rothamsted would not grow any crop for several years.

Herbicides are usually more water-soluble than insecticides and, hence, are more readily leached into the sub-soil. Fenac and 2,3,6-trichlorobenzoic acid, which are more water-soluble and more resistant to microbial breakdown than most herbicides, have been found at depths of two to five feet below the soil surface as long as three years after they were applied.[305 313] Herbicides which are less soluble, such as atrazine, have also been found below the cultivation layer; and it has been suggested that microbial breakdown is slower in the subsoil than in the surface soil.[301] This has not been proved and seems unlikely because many of the microorganisms important in degrading herbicides are anaerobic.

In recent years, organic arsenical compounds have been used as contact herbicides, and these may break down to compounds which persist in soil longer than other herbicides or their residues. Some herbicides, such as sodium cacodylate and the sodium salts of methanearsonic acid, are slowly metabolized by soil microorganisms to arsenate and carbon dioxide. These arsenic residues are very persistent in soil and may kill soil microorganisms or damage crops.[150] However, it has been calculated[316] that when disodium methanearsonic acid has been used to control weeds in cotton in the southern United States, even after annual applications of 6 kg per ha, that it would take 50 years before the arsenic levels in the soil affected the growth of cotton. Other crops, such as rice, are much more susceptible to arsenic residues in the soil, and the soil fauna may also be seriously affected by arsenic residues.

Most other herbicides do not affect numbers of soil animals very much. The only ones that have been reported to kill soil animals are DNOC and simazine.[75 77] The soil microorganisms are not affected by the great majority of herbicides if the carbon dioxide output of soil is used as an index of microbial activity.[308 309] Even when herbicides cause a reduction in respiration, this is usually temporary and is followed by an increase in microbial activity. The nitrifying bacteria seem to be among the more sensitive of the soil microorganisms to herbicides.

Some herbicides can persist for several months in lake water and lake-bottom muds; the sodium salt of 2,4-D persisted for 120 days in aerated natural lake waters,[306] and there have been other reports of this herbicide persisting in water for 3 to 8 years,[307] although not usually in potentially harmful quantities. 2,6-dichlorobenzonitrile has been reported to persist for 188 days in a farm pond, and some was taken up into fish.[319] In general, however, herbicides do not persist for long periods in water, so any hazard they cause is an immediate one rather than long-term. Most herbicides are less toxic to fish than are insecticides, although at concentrations for "effective" control of aquatic weeds, fish populations are often affected.[311 312] The most potentially harmful herbicides to fish seem to be 2,4-D and MCPA, and the most common histological damage in poisoned fish is to the liver. DNOC and pentachlorophenol are also toxic, and simazine, diquat and paraquat moderately toxic to fish.

Recently, there has been some evidence that 2,4,5-T can cause harm to animal life. The hazard connected with the use of this herbicide seems to be that it can cause reproductive disturbances, abortions and the birth of deformed offspring in vertebrates. This is the subject of further research but, in the interim, use of this chemical for aerial spraying has been restricted in the United States and for some purposes in Great Britain.

In general, the herbicides do not seem to be very hazardous to wild animals although their indiscriminate use may destroy many natural habitats.

Fungicides

Most of the organic fungicides are biodegradable and persist in soil for only a few days or weeks; one of the most persistent is quintozene which takes several months to break down. Most of these chemicals have low mammalian toxicity, they are not taken up into biological material and are usually used only in small quantities. Of the fungicides, the organic mercury compounds are the most potentially harmful to wild life, but these are most commonly used as seed treatments so that only small quantities reach the soil. However, the treated seed may be a hazard to birds which feed on it, particularly because mercury persists in animal tissues.

The fungicides which persist longest are the inorganic ones which contain heavy metals. Many of these break down to leave residues which contain copper and mercury, which are very persistent in soils. These copper and mercury residues may be harmful to the soil fauna and microflora; for instance, very large residues of copper exist in many orchards after regular spraying with 'Bordeaux mixture.' In a survey of orchards in southeast England, those orchards with a long history of heavy spraying with copper fungicides had very few earthworms, and the soil structure was very poor.[314] Copper residues can also harm crops growing in contaminated soils, causing damage such as copper chlorosis.

There are few data on the hazards of fungicides to soil animals, but there is little doubt that fungicides would affect the soil microflora if they persisted for long in soil; fortunately, not many fungicides are used for soil treatments, and those that are do not persist for long. Little reliable information exists on the toxicity of the newer organic fungicides to fish and other aquatic organisms. However, the older inorganic fungicides such as copper sulfate, mercuric chloride and phenylmercuric acetate are very toxic to fish.

Residues of mercury have been reported in many wild animals, sometimes in large amounts. For instance, Borg et al.[299] found considerable amounts (up to 140 mg/kg) of mercury, in seed-eating birds and their predators, and even in small rodents. The amounts were large enough to have been responsible for the deaths of some of these animals. When partridges were fed experimentally on seed dressed with Thiram, their egg-laying was significantly decreased.[310] Residues of the fungicide hexachlorobenzene have been found in the tissues of wild birds in Holland;[320] presumably these came from eating seed that had been dressed with this fungicide.

There are many other isolated examples of fungicides harming wild animals but there is little evidence that fungicides cause much damage to the environment compared with the persistent insecticides.

Occurrence of Resistance to Persistent Pesticides

Continued exposure to a particular insecticide often results in a pest becoming resistant to that insecticide. In view of the large quantities of chlorinated hydrocarbon insecticides that have been used in all parts of the world, it is not surprising that many pests have developed resistance to these chemicals. To control a pest that has become resistant to an insecticide, either much more of the same insecticide is used or alternative insecticides are necessary. If the same insecticide is used, the amounts required to control the pest adequately may continue to increase as further resistance develops.[302] On a world scale, this means that the large amounts of persistent pesticides already being used would continue to increase. Thus it is important to assess the current position as to how much resistance to the persistent pesticides currently exists in various parts of the world.

Resistance to the chlorinated hydrocarbon insecticides is of two main types; firstly, resistance to DDT and its analogues, and secondly to the cyclodiene derivatives such as dieldrin, aldrin and chlordane. Resistance to the cyclodiene insecticides is often associated with resistance to gamma benzene hexachloride. The degree of linking of resistance to a particular pesticide with resistance to other pesticides varies considerably. Sometimes the resistance may be completely specific to a particular insecticide, but, more commonly, resistance to one insecticide also involves resistance to one or more other insecticides. Resistance to chlorinated hydrocarbon insecticides may even be associated with resistance to organophosphate insecticides; this has been demonstrated with both houseflies and culicine mosquitoes.

Not all resistance can be overcome by increasing the dose of an insecticide; for instance, when resistance to dieldrin occurs, larger doses are usually ineffective. Conversely, when anopheline mosquitoes become resistant to DDT, the transmission of malaria may still be prevented by more frequent applications of this insecticide.

Resistance to chlorinated hydrocarbon insecticides may be confined to local areas; for instance, DDT-resistant codling moths have been limited to just a few orchards for periods of several years. Or the resistance may spread rapidly; for example, resistance of the western corn rootworm to aldrin spread from a single spot in Nebraska to seven other states in five years. Dieldrin resistance usually develops very rapidly; for example, the cattle tick (*Boophilus microplus*) in Australia and *Anopheles gambiae* in Nigeria became resistant within six months of dieldrin being used. By contrast, DDT-resistance often takes a long time to develop; it took two years in the housefly in Mediterranean climates, and in temperate countries it took five years for DDT-resistance to develop in codling moth (*Laspeyresia pomonella*) and cabbage caterpillars.

It seems clear that the surest way of inducing resistance in pests is to contaminate as wide an area as possible with persistent insecticides, which kill a large proportion of the insects in the area, for a considerable period.[300] The boll weevil (*Anthonomus grandis*) in southern U.S.A., the cotton leafworm (*Spodoptera littoralis*) in the Nile Delta, and mosquitoes in interior California and coastal Florida are good examples of this principle.

Because chlorinated insecticides have been so widely used, it is reassuring that, although many pests have become resistant to these chemicals (Table 23), there are not many pests where resistance to chlorinated hydrocarbon insecticides has become so widespread that alternative methods of control are essential.[302] Among the pests where this has become necessary are the mosquito vector of Bancroftian filiariasis (*Culex fatigans*), the housefly (*Musca domestica*), the cotton boll weevil (*Anthonomus grandis*), the sheep blowfly (*Lucilia cuprina*), the cabbage maggot (*Hylemya brassicae*), Colorado beetle (*Leptinotarsa decemlineata*), northern corn rootworm (*Diabrotica longicornis*), alfalfa weevil (*Hypera postica*) the cotton bollworms and the tobacco hornworm (*Protoparce sexta*). There are many other pests where resistance has developed in some geographical areas and which may eventually pose serious problems if the resistance continues to spread.

TABLE 23

Numbers of Species of Arthropods with Populations Resistant to Chlorinated Hydrocarbon Insecticides

Arthropod group	DDT	Cyclodiene insecticides
Diptera	44	68
Lepidoptera	14	14
Hemiptera	10	15
Coleoptera	5	19
Acarina	3	7
Other orders	15	12
Totals	91	135

From Brown, A.W.A.[300]

It seems that the problem of resistance to persistent insecticides is not yet as serious as it might be. Of some 5,000 species of insects and mites that are of economic importance as pests or disease vectors, only about 5% have so far developed resistance to the chlorinated hydrocarbon insecticides. Unfortunately, this 5% includes some of the more economically important species, particularly the Diptera which transmit diseases, so there is still need for some anxiety as to the importance of resistance. The possibility that if chlorinated hydrocarbon insecticides continue to be used on a large scale, then resistance will continue

to increase as well, is an important factor in making decisions as to the future use of the pesticides.

Concentration of Chlorinated Pesticides or Their Metabolites in Human Milk

Table 24 summarizes some data on the levels of chlorinated pesticides or their metabolites found in human milk. The highest concentrations were reported for DDT and DDE.

Common and Chemical Names of Pesticides Referred to in Text

Aldrin 1,2,3,4,10,10-hexachloro-1,4,4a,5,8,8a,-hexahydro-*exo*-1,4-*endo*-5,8-dimethanonaphthalene

Atrazine 2-chloro-6-ethylamino-4-isopropylamino-1,-3,5-triazine

Benzene hexachloride Mixed isomers of 1,2,3,4,5,6-hexachlorocyclohexane

Bordeaux mixture copper sulphate (with a lime suspension)

Chlordane 1,2,4,5,6,7,10,10-octachloro-4,7,8,9-tetrahydro-4,7-methyleindane

2,4-D 2,4-dichlorophenoxyacetic acid

Dichlobenil 2,6-dichlorobenzonitrile

DDD *See* TDE

DDT 1,1,1-trichloro-2,2-bis(*p*-chlorophenyl)ethane

Dieldrin 1,2,3,4,10,10-hexachloro-6,7-expoxyl-1,4,4a,5,-6,7,8,8a-octahydro-*exo*-1,4-*endo*-5,8-dimethanonaphthalene

DNOC 2-methyl-4,6-dinitrophenyl

Disodium methanearsonic acid ————

Diuron *N'*-(3,4-dichlorophenyl)-*NN*-dimethyl urea

Endosulfan 6,7,8,9,10,10-hexachloro-1,5,5a,6,9,9a-hexahydro-6,9-methano-2,4,3-benzo(*e*)-dioxathiepin 3-oxide

Endrin 1,2,3,4,10,10-hexachloro-6,7-epoxy-1,4,4a,5,-6,7,8,8a-octahydro-*exo*-1,4-*exo*-5,8-dimethanonaphthalene

Fenac 2,3,6-trichlorophenylacetic acid

HEOD *See* dieldrin

Heptachlor 1,4,5,6,7,10,10-heptachloro-4,7,8,9-tetrahydro-4,7-methyleindene

Hexachlorobenzene ————

Lindane *See* benzene hexachloride

Mercuric chloride ————

Monuron *N'*-4-chlorphenyl)-*NN*-dimethylurea

Neburon *N*-butyl-*N'*-(3,4-dichlorophenyl)-*N*-methylurea

Parathion diethyl 4-nitrophenyl phosphorthionate

Pentachlorophenyl ————

Phenyl mercuric acetate ————

Picloram 4-amino-3,5,6-trichloropicolinic acid

Propazine 2-chloro-4,6-bisisopropylamino-1,3,5-triazine

Propham   isopropyl *N*-phenylcarbamate
Rhothane   *See* TDE
Simazine   2-chloro-4,6-bisethylamino-1,3,5-triazine
Sodium cacodylate   ————
TDE   1,1-dichloro-2,2-bis(*p*-chlorophenyl)ethane
2,4,5-T   2,4,5-trichlorophenoxyacetic acid
Thiram   bis(dimethylthiocarbamoyl)disulphide
Toxaphene   chlorinated camphene containing 67–
   69% chlorine

Conversion Factors

kg/ha	x 0.892	=	lb/acre
tons	x 0.907	=	metric tons
acres	x 0.4047	=	hectares
lb	x 0.4536	=	kilograms
lb/acre	x 1.120	=	kg/ha
mile	x 1.609	=	kilometers

TABLE 24

Concentration of Chlorinated Pesticides or Their Metabolites in Human Milk

Pesticide or metabolite	Country	Year	No. of samples from non-exposed people	Analysis method**	Concentration (ppm)		Ref. (see below)
DDT	U.S.A.	1950	32	Col.	0.13		23
p,p' + *o,p'*	U.S.A.	1960–61	10	Col.	0.08		24
	U.S.A.	1962	6	Col.	0–0.12		25
	Hungary	1963	10	Col.	0.13–0.26		26*
	England	1963–64	19	G.L.C.	0.05		13
	Belgium	1966	20	G.L.C.	0.046		27
p,p'-DDE	U.S.A.	1960–61	10	Col.	0.036	0.04†	24
as DDT	U.S.A.	1962	6	Col.	0–0.025	0–0.028†	25
	England	1963–64	19	G.L.C.	0.073	0.08†	13
	Belgium	1966	20	G.L.C.	0.072	0.08†	27
Total BHC	England	1963–64	19	G.L.C.	0.013		13
BHC Isomers	Belgium	1965–66	20	G.L.C.	0.010(γ)		27
Dieldrin	England	1963–64	19	G.L.C.	0.006		13
	Belgium	1965–66	20	G.L.C.	0.003		27

* Based on 4% fat.
** Col.—colormetric; G.L.C.—Gas-liquid chromatography.
† *p,p'-DDE* as DDT.
  Adapted from Table 22, Concentration of Chlorinated Pesticides or Their Metabolites in Human Material, in Sunshine, I., *Handbook of Analytical Toxicology,* Chemical Rubber Co., Cleveland, 1969, 566.

References for Table 24

13. Egan, H. et al., *Brit. Med. J.,* 2, 66, 1965.

23. Laug, E.P., Kunze, F.M., and Prickett, C.S., *Arch. Ind. Hyg. Occupational Med.,* 3, 245, 1951.

24. Quinby, G.E., Armstrong, J.E., and Durham, W.F., *Nature,* 207, 726, 1965.

25. West, I., *Arch. Environ. Health,* 9, 626, 1964.

26. Denes, A., *1962 Yearbook,* Inst. Nutr., Budapest, 1963, 47.

27. Maes, R.A. and Heyndrickx, A., *J. Pharm. Belg.,* 1968.

REFERENCES

1. Abbott, D.C., Harrison, R.B., Tatton. J. O'G., and Thompson, J., Organochlorine insecticides in the atmospheric environment, *Nature,* London, 211, 259, 1966.

2. Abbott, D.C., Holmes, D.C., and Tatton, J. O'G., Pesticide residues in the total diet in England and Wales 1966-7, *J. Sci. Food Agr.,* 20, 242, 1969.

3. Adkisson, P.L. and Wellso, S.G., Effect of DDT poisoning on the longevity and fecundity of the pink bollworm, *J. Econ. Entomol.,* 55, 842, 1962.

4. Allen, N., Walker, R.L., and Fife, L.C., Persistence of BHC, DDT and toxaphene in soil and the tolerance of certain crops to their residues, *Tech. Bull. U.S. Dept. Agr.* 1090, 1954, 19.

5. Ames, P.L., DDT residues in the eggs of the osprey in the North-Eastern United States and their relation to nesting success, *J. Appl. Ecol. (Suppl.),* 3, 87, 1966.

6. Anon., Pesticides and their effects on soils and waters, *Soil Science Society of America*, 1966, 150.

7. Arnason, A.P., Brown, A.W.A., Fredeen, F.J.H., Hopewell, W.W., and Rempel, J.G., Experiments in the control of *Simulium articum* Mall by means of DDT in the Saskatchewan river, *Sci. Agr.,* 29, 527, 1951.

8. Azevedo, J.A. Jr., Hunt, E.G., and Woods, L.A. Jr., Physiological effects of DDT on pheasants, *Calif. Fish Game,* 51, 276, 1965.

9. Bailey, G.W. and White, J.L., Review of adsorption and desorption of organic pesticides by soil colloids, with implications concerning pesticide bioactivity, *J. Agr. Food Chem.,* 12, 324, 1964.

10. Barker, R.J., Notes on some ecological effects of DDT sprayed in elms, *J. Wildlife Manage.,* 22(3), 269, 1958.

11. Barlow, F. and Hadaway, A.B., Studies on aqueous suspensions of insecticides. Further note on the sorption of insecticides in soils, *Bull. Entomol. Research,* 49, 315, 1958.

12. Barney, J.E., Pesticide pollution of the air studied, *Chem. Eng. News,* 42, Jan. 1969.

13. Beard, R., Ovarian suppression by DDT and resistance in the housefly (*Musca domestica* L.), *Entomol. Exp. Appl.,* 8, 193, 1965.

14. Beestman, G.B., Keeney, D.R. and Chesters, K., Dieldrin translocation and accumulation in corn, *Agron. J.,* 61, 390, 1969.

15. Bernard, R.F., Studies on the effects of DDT on birds, *Publs. Mich. State University Museum,* 2, 155, 1963.

16. Bernard, R.F. and Gaertner, R.A., Some effects of DDT on reproduction in mice, *J. Mamm.,* 45(2), 272, 1963.

17. Bick, M., Chlorinated hydrocarbon residues in human body fat, *Med. J. Austral.,* 1, 1127, 1967.

18. Bowman, M.C., Schechter, M.S., and Carter, R.L., Behaviour of chlorinated insecticides in a broad spectrum of soil types, *J. Agr. Food Chem.,* 13, 360, 1965.

19. Brady, N.C., Agriculture and the Quality of Our Environment, *American Association for the Advancement of Science,* 1966, 460.

20. Breidenbach, A.W., Pesticide residues in air and water, *Arch. Environ. Health,* 10, 827, 1965.

21. Breidenbach, W.W., Gunnerson, C.G., Kawahara, F.K., Lichtenberg, J.J., and Green, R.S., Chlorinated hydrocarbon pesticides in major river basins 1957-65, *Publ. Health Rept. Wash.,* 82, 139, 1967.

22. Brett, C.H. and Bowery, T.G., Insecticide residues on vegetables, *J. Econ. Entomol.,* 51, 818, 1958.

23. Brewerton, H.V., and McGrath, H.J.W., Insecticides in human fat in New Zealand, *N.Z.J. Sci.,* 10, 486, 1967.

24. Bridges, W.R., Kallman, B.J., and Andrews, A.K., Persistence of DDT and its metabolites in a farm pond, *Trans. Amer. Fish Soc.,* 92, 421, 1963.

25. Brown, W.A. et al., Secretion of DDT in fresh milk by cows, *Bull. Environ. Contam. Toxicol.,* 1, 21, 1966.

26. Brown, E. and Nishioka, Y.A., Pesticides in selected western streams—a contribution to the National Program, *Pest. Mon. J.,* 1(2), 38, 1967.

27. Brown, J.R., Organochlorine pesticide residues in human depot fat, *Can. Med. Ass. J.*, 97, 367, 1967.

28. Bruce, W.N., Decker, G.C., and Wilson, J.G., The relationship of the levels of insecticide contamination of crop seeds to the fat content and soil concentration of aldrin, heptachlor and their epoxides, *J. Econ. Entomol.*, 59, 179, 1966.

29. Burdick, G.E., Harris, E.J., Dean, H.J., Walker, T.M., Shed, J., and Colby, D., The accumulation of DDT in lake trout and the effect on reproduction, *Trans. Amer. Fish. Soc.*, 93, 127, 1964.

30. Burrage, R.H. and Saha, J.G., Insecticide residues in spring wheat plants grown in the field from seed treated with aldrin or heptachlor, *Can. J. Plant Sci.*, 47, 114, 1967.

31. Butcher, A.D., Wildlife hazards from the use of pesticides, *Austral. J. Pharmacol.*, 105, 1965.

32. Butler, P.A., Commercial fishery investigations in *Effects of Pesticides on Fish and Wildlife, U.S.D.I. Fish Wildlife Circ.*, 226, 65, 1965.

33. Butler, P.A., Pesticides in the marine environment, *J. Appl. Ecol. (Suppl)*, 3, 253, 1966.

34. Cain, S.A., Pesticides in the environment with special attention to aquatic biology resources, *Rept. U.S. Japan Meeting on Pesticide Research*, 12, 1965.

35. Campbell, J.E., Richardson, L.A., and Schafer, M.L., Insecticide residues in the human diet, *Arch. Environ. Health*, 10, 831, 1965.

36. Carollo, J.A., The removal of DDT from water supplies, *J. Amer. Water Works Ass.*, 37, 1310, 1945.

37. Carson, Rachel, *Silent Spring*, Houghton-Mifflin Co., Boston, 1962, 368.

38. Cassidy, W., Fisher, A.J., Peden, J.D., and Parry-Jones, A., Organochlorine pesticide residues in human fats from Somerset, *Monthly Bull. Ministr. Health London*, 26, 1, 1967.

39. Chichester, C.O., *Research in Pesticides*, Academic Press, New York, 1965, 380.

40. Chisholm, R.D., Koblitsky, L., Fahey, J.E., and Westlake, W.E., DDT residues in soil, *J. Econ. Entomol.*, 43(6), 941, 1951.

41. Chisholm, D., Koblitsky, L.K., Fahey, J.E., and Westlake, W.E., DDT residues in soil, *Trans. 24th N. Amer. Wildlife Conf.*, 118, 1959.

42. Cohen, J.M., Kamphake, L.J., Lempke, A.E., Henderson, C., and Woodward, R.L., Effect of fish poisons on water supplies Part 1, Removal of toxic materials, *J. Amer. Water Works Ass.*, 52 (12), 1551, 1960.

43. Cohen, J.M. and Pinkerton, C., Widespread translocation of pesticides by air transport and rain-out, *Adv. Chem. Ser.*, 60, 163, 1966.

44. Cole, H., Barry, D., Frear, D.E.H., and Bradford, A., DDT levels in fish, streams, stream sediments and soil before and after DDT aerial spray application for fall cankerworm in northern Pennsylvania, *Bull. Environ. Contam. Toxicol.*, 2, 127, 1967.

45. Cope, O.B., in Sport fisheries investigations—Pesticide-Wildlife studies 1963, *USDI Fish Wildlife Serv. Circ.*, 199, 29, 1963.

46. Cope, O.B., Contamination of the freshwater ecosystem by pesticides, *J. Appl. Ecol. (Suppl)*, 3, 33, 1966.

47. Corneliussen, P.E., Pesticide residues in total diet samples, *Pest. Mon. J.*, 2(4), 140, 1969.

48. Cramp, S. and Conder, P.J., Fifth Report of the Joint Committee of the British Trust for Ornithology, *Roy. Soc. Prot. Birds Rept.*, 1965, 20.

49. Cramp, S. and Olney, P.J.S., Sixth report of the Joint Committee of the British Trust for Ornithology, *Roy. Soc. Prot. Birds Rept.*, 1966, 26.

50. Cramp, S., Conder, P.J., and Ash, J.S., The risk to bird life from chlorinated hydrocarbon pesticides, *Roy. Soc. Prot. Birds Rept.*, 1964, 24.

51. Crocker, R.A. and Wilson, A.J., Kinetics and effects of DDT in a tidal marsh ditch, *Trans. Amer. Fish Soc.*, 94(2), 152, 1965.

52. Cummings, J.G., Eidelman, M., Turner, V., Reed, D., Zee, K.T., and Cock, R.E., Residues in poultry tissues from low level feeding of five chlorinated hydrocarbon insecticides to hens, *J. Ass. Off. Anal. Chem.*, 50(2), 418, 1967.

53. Dale, W.E. and Quinby, G.E., Chlorinated insecticides in the body fat of people in the United States, *Science (NY)*, 142, 593, 1965.

54. Dale, W.E., Gaines, T.B., and Hayes, W.T. Jr., Chlorinated insecticides in the body fat of people in India, *Bull. World Health Org.*, 33, 471, 1965.

55. Davis, B.N.K., The soil macrofauna and organochlorine insecticide residues at twelve agricultural sites near Huntingdon, *Ann. Appl. Biol.*, 61, 29, 1968.

56. Davis, B.N.K. and French, M.C., The accumulation and loss of organochlorine insecticide residues by beetles, worms and slugs in sprayed fields, *Soil Biol. Biochem.*, 1, 45, 1969.

57. Davis, B.N.K. and Harrison, R.B., Organochlorine insecticide residues in soil invertebrates, *Nature,* London, 211, 1424, 1966.

58. David, K.B., The effects of lethal and sublethal doses of DDT on the learning ability group behaviour and liver glycogen storage in the bobwhite *(Colinus virginianus)*, Thesis, Univ. of Ark., 1965.

59. Decker, G.C., Bruce, N.W., and Bigger, J.H., The accumulation and dissipation of residues resulting from the use of aldrin in soils, *J. Econ. Entomol.*, 58(2), 266, 1965.

60. Dennis, E.B. and Edwards, C.A., Phytotoxicity of insecticides and acaricides. III. Soil applications, *Plant Pathol.*, 13(4), 173, 1964.

61. Deubert, K.H. and Zuckerman, B.M., Distribution of dieldrin and DDT in cranberry bog soil, *Pest. Mon. J.*, 2(4), 172, 1969.

62. DeWitt, J.B., Chronic toxicity to quail and pheasants of some chlorinated insecticides, *J. Agr. Food Chem.*, 4, 863, 1956.

63. DeWitt, J.B., Crabtree, D.G., Finley, R.B., and George, J.L., Effects of pesticides on fish and wildlife in 1960. Effects on wildlife-A review of investigations during 1960. *U.S.D.I. Fish Wildlife Circ.*, 143, 4, 1962.

64. DeWitt, J.B., Menzie, C.M., Adomaitis, V.A., and Reichel, W.L., Pesticidal residues in animal tissues, *Trans. 25th N. Amer. Wildlife Conf.*, 277, 1960.

65. Doane, C.C., Effects of certain insecticides on earthworms, *J. Econ. Entomol.*, 55(3), 416, 1962.

66. Dormal, S., Martens, P.H., Decleire, M., and de Faestraets, L., Study of the persistence of insecticide residues in various vegetables, fresh, blanched and sterilized, *Bull. Inst. Agron. Sta. Rech. Gemboux*, 27, 137, 1959.

67. Duffy, J.R. and O'Connell, D., DDT residues and metabolites in Canadian Atlantic coast fish, *J. Fish Res. Bd. Can.*, 25, 189, 1968.

68. Duffy, J.R. and Wong, N., Residues of organo-chlorine insecticides and their metabolites in soils in the Atlantic provinces of Canada, *J. Agr. Food Chem.*, 15, 457, 1967.

69. Duggan, R.E., Pesticide residue levels in foods in the United States from July 1, 1963 to June 30, 1967, *Pest. Mon. J.*, 2(1), 2, 1968.

70. Dustan, G.G. and Chisholm D., DDT residues on peach in Ontario, *J. Econ. Entomol.*, 52, 109, 1959.

71. Dustman, E.H. and Stickel, L.F., Pesticide residues in the ecosystem, in *Pesticides and Their Effects on Soils and Water,* Amer. Soc. Agron. Spec. Publ., 8, 109, 1966.

72. Dustman, E.H. and Stickel, L.F., The occurrence and significance of pesticide residues in wild animals, Conference on Biological Effects of Pesticides in Mammalian Systems, *N.Y. Acad. Sci.*, (in press)

73. Eades, J.F., Pesticide residues in the Irish environment, *Nature,* London, 210, 650, 1966.

74. Edmundson, W.F., Davies, J.E., and Hull, W., Dieldrin storage levels in necropsy adipose tissue from a South Florida population, *Pest. Mon. J.*, 2(2), 86, 1968.

75. Edwards, C.A., Effects of pesticide residues on soil invertebrates and plants, *Ecology and the Industrial Society,* Blackwell, Oxford, 1965, 239.

76. Edwards, C.A., Insecticide residues in soils, *Residue Reviews*, 13, 83, 1966.

77. Edwards, C.A., Soil pollutants and soil animals, *Sci. Amer.*, 220(4), 88, 1969.

78. Edwards, C.A., Loss of aldrin and dieldrin residues from soil by leaching and ecological studies of their residues, *Aldrin Dieldrin Symp. Washington*, 1, 7, 1969.

79. Edwards, C.A., The problem of insecticidal residues in agricultural soils, *N.A.A.S. Quart. Rev.*, 86, 47, 1969.

80. Edwards, C.A., Beck, S.D., and Lichtenstein, E.P., Bioassay of aldrin and lindane in soil, *J. Econ. Entomol.*, 50, 622, 1957.

81. Edwards, C.A., Dennis, E.B., and Empson, D.W., Pesticides and the soil fauna. I. Effects of aldrin and DDT in an arable field, *Ann. Appl. Biol.*, 59(3), 11, 1967.

82. Edwards, C.A., Thompson, A.R., and Lofty, J.R., Changes in soil invertebrate populations due to some organophosphorus insecticides, *Proc. 4th Brit. Ins. Fung. Conf.*, 48, 1967.

83. Edwards, C.A., Thompson, A.R., and Beynon, K.I., Some effects of chlorfenvinphos, an organophosphorus insecticide on populations of soil animals, *Rev. Ecol. Biol. Sol.*, 5(2), 199, 1967.

84. Edwards, C.A., Thompson, A.R., Beynon, K.I., and Edwards, M.J., Movement of dieldrin in soils. I. From arable soils into ponds, *J. Sci. Food Agr.*, (in press)

85. Edwards, R.W., Egan, H., Learner, M.A., and Maris, P.J., The control of chironomid larvae in ponds using TDE (DDD), *J. Appl. Ecol. (Suppl)*, 3, 97, 1966.

86. Egan, H., Organochlorine pesticide residues in human fat and human milk, *Brit. Med. J.*, 2, 66, 1965.

87. Ely, R.E., Moore, L.A., Carter, R.H., and App, B.A., Excretion of endrin in the milk of cows fed endrin - sprayed alfalfa and technical endrin, *J. Econ. Entomol.*, 50, 348, 1957.

88. Epstein, E. and Grant, W.J., Chlorinated insecticides in run off water as affected by crop rotation, *Proc. Soil Sci. Amer.*, 32(3), 423, 1968.

89. Fahey, J.E., Butcher, J.W., and Murphy, R.T., Chlorinated hydrocarbon insecticide residues in soils of urban areas in Battle Creek, Michigan, *J. Econ. Entomol.*, 58(5), 1026, 1965.

90. Fahey, J.E., Redriguez, J.G., Rusk, H.W., and Chaplin, C.E., Chemical evaluation of pesticide residues on strawberries, *J. Econ. Entomol.*, 55, 179, 1962.

91. Ferguson, D.E., Culley, D.D., Colton, W.D., and Dodds, R.P., Resistance to chlorinated hydrocarbon insecticides in 3 spp of freshwater fish, *Bio. Sci.*, 14, 43, 1965.

92. Finlayson, D.G. and McCarthy, H.R., The movement and persistence of insecticides in plant tissues, *Residue Reviews*, 9, 114, 1965.

93. Fiserova, B.V., Radonski, J.L., Davies, J.E., and Davis, J.H., Levels of chlorinated hydrocarbon pesticides in human tissues, *Ind. Med. Surg.*, 36, 65, 1967.

94. Fleck, E.E. and Haller, H.L., Compatability of DDT with insecticides, fungicides and fertilizers, *Ind. Eng. Chem. Eng. Anal. Ed.*, 37, 403, 1945.

95. Fleming, W.E., Bioassay of soil containing residues of chlorinated hydrocarbon insecticides with special reference to control of Japanese beetle (*P. japonica*) grubs, *U.S.D.A. Tech. Bull.*, 1266, 1962, 44.

96. Fleming, W.E. and Maines, W.W., Persistence of DDT in soils of the area infested by the Japanese beetle, *J. Econ. Entomol.*, 46(3), 445, 1953.

97. Foster, A.C., Some plant responses to certain insecticides in the soil, *U.S.D.A. Circ.*, 862, 1951, 41.

98. Fowkes, F.M., Benes, H.A., Ryland, L.B., Sawyer, W.M., Detling, K.D., Loeffler, E.S., Folckemer, F.B., Johnson, M.R., and Sun, Y.P., Clay-catalyzed decomposition of insecticides, *J. Agr. Food Chem.*, 8, 203, 1960.

99. Gallaher, P.J. and Evans, L., Preliminary investigations on the penetration and persistence of DDT under pasture, *N.Z. J. Agr. Res.*, 4, 466, 1961.

100. George, J.L. and Frear, D.E.H., Pesticides in the Antarctic, *J. Appl. Ecol. (Suppl.)*, 3, 155, 1966.

101. Getzin, L.W. and Rosefield, I., Organophosphorus insecticide degradation by heat - labile substances in soil, *J. Agr. Food Chem.*, 16(4), 598, 1968.

102. Ginsburg, J.M. and Reed, J.P., A survey of DDT accumulation in soils in relation to different crops, *J. Econ. Entomol.*, 47(3), 467, 1954.

103. Godsil, P.J. and Johnson, W.C., Pesticide monitoring of the aquatic biota at the Tule Lake National Wildlife Refuge, *Pest. Mon. J.*, 1(4), 21, 1968.

104. Gorham, J.R., Aquatic insects and DDT forest spraying in Maine, *Maine For. Serv. Conserv. Found. Bull.*, 19, 1, 1961.

105. Gould, R.F., *Organic Pesticides in the Environment*, American Chemical Society, 1966, 309.

106. Green, R.S., Gunnerson, G.C., and Lichtenberg, J.J., Pesticides in our national waters, in *Agriculture and the Quality of our Environment*, American Association for the Advancement of Science, 1966, 137.

107. Guenzi, W.D. and Beard, W.E., Stability of DDT in soil under aerobic and anaerobic conditions, *Proc. Soil Soc. Amer.*, 32(4), 522, 1968.

108. Gyrisco, G.G., Norton, L.B., Trimberger, G.W., Holland, R.F., and McEnerney, P.J., Effects of feeding low levels of insecticide residues on hay to dairy cattle on flavor and residues in milk, *J. Agr. Food Chem.*, 7, 707, 1959.

109. Hamaker, J.W., Mathematical prediction of cumulative levels of pesticides in soil, in *Organic Pesticides in the Environment, Amer. Chem. Soc. Adv. Chem. Ser.*, 60, 122, 1966.

110. Hansen, D.J., Indicator organisms - fish, *Ann. Rept. Bur. Comm. Fish Circ.*, 247, 10, 1966.

111. Hardee, D.D., Huddleston, E.W., and Gyrisco, G.C., Initial deposits and disappearance rates of various insecticides as affected by forage crop species, *J. Econ. Entomol.*, 56. 98, 1963.

112. Harris, C.R., Factors affecting the volatilization of insecticides from soils, *Diss. Abstr.*, 22, 375, 1961.

113. Harris, C.R., Influence of soil moisture on the toxicity of insecticides in a mineral soil to insects, *J. Econ. Entomol.*, 57, 946, 1964.

114. Harris, C.R., Influence of soil type on the activity of insecticides in soil, *J. Econ. Entomol.*, 59(5), 1221, 1966.

115. Harris, C.R. and Lichtenstein, E.P., Factors affecting the volatilization of insecticides from soils, *J. Econ. Entomol.*, 54, 1038, 1961.

116. Harris, C.R., Sans, W.W., and Miles, J.R.W., Exploratory studies on occurrence of organochlorine residues in agricultural soils in S.W. Ontario, *J. Agr. Food Chem.*, 14, 398, 1966.

117. Harvey, J.M., Excretion of DDT by migratory birds, *Can. J. Zool.*, 45, 629, 1967.

118. Hayes, W.J. Jr., Review of the metabolism of chlorinated hydrocarbon insecticides, especially in mammals, *Ann. Rev. Pharmacol.*, 5, 27, 1965.

119. Hayes, W.J., Monitoring food and people for pesticide content, in *Scientific Aspects of Pest Control*, Nat. Acad. Sci. Pub., 1402, 314, 1966.

120. Hayes, W.J. and Curley, A., Storage and excretion of dieldrin and related compounds, *Arch. Environ. Health*, 16, 155, 1968.

121. Hayes, W.J., Dale, W.E., and Le Breton, R., Storage of insecticides in French people, *Nature*, London, 199, 1189, 1963.

122. Hayes, W.J. Jr., Dale, W.E., and Birse, V.W., Chlorinated hydrocarbon pesticides in the fat of people in New Orleans, *Life Sci.*, 4, 1611, 1965.

123. Hayes, W.J. Jr., Quinby, G.E., Walker, K.C., Elliott, J.W., and Upholt, W.M., Storage of DDT and DDE in people with different degrees of exposure to DDT, *Arch. Ind. Health*, 18, 398, 1958.

124. Hermanson, H.P. and Forbes, C., Soil properties affecting dieldrin toxicity to *Drosophila melanogaster, Proc. Soil Sci. Soc. Amer.*, 30, 748, 1966.

125. Hickey, J.J., DDT and birds: Wisconsin 1968., *Atlantic Nat.*, 24(2), 86, 1969.

126. Hickey, J.J. and Anderson, D.W., Chlorinated hydrocarbons and egg shell changes in raptorial and fish-eating birds, *Science N.Y.*, 162, 271, 1968.

127. Hickey, J.J., Keith, J.A., and Coon, F.B., An exploration of pesticides in a Lake Michigan ecosystem, *J. Appl. Ecol. (Suppl)*, 3, 141, 1966.

128. Hitchcock, S.W., Field and laboratory studies of DDT and aquatic insects, *Conn. Agr. Exp. Sta. Bull.*, 668, 1965, 32.

129. Hoffman, W.S., Fishbein, W.I., and Andelman, M.B., The pesticide content of human fat tissue, *Arch. Environ. Health*, 9, 387, 1964.

130. Holden, A.V., Organochlorine insecticide residues in salmonid fish, *J. Appl. Ecol. (Suppl)*, 3, 45, 1966.

131. Holden, A.V. and Marsden, K., Examination of surface waters and sewage effluents for organo-chlorine pesticides, *J. Proc. Sew. Purif.*, 4, 295, 1966.

132. Holden, A.V. and Marsden, K., Organochlorine pesticides in seals and porpoises, *Nature*, London, 216(5122), 1274, 1967.

133. Howell, D.E., A case of DDT storage in human fat, *Proc. Okl. Acad. Sci.*, 29, 31, 1948.

134. Hunt, E.G., Pesticide residue studies, *Fed. Aid Proj. FW-1-RI Job Compl. Rept. WP-1*, J-2 1963 - 64, 1964, 12.

135. Hunt, E.G. and Bischoff, A.I., Inimical effects on wildlife of periodic DDD applications to Clear Lake, *Calif. Fish Game*, 46, 91, 1960.

136. Hunt, L.B., Kinetics of pesticide poisoning in Dutch Elm Disease control. The effects of pesticides on fish and wildlife, *U.S.D.A. Wildlife Serv. Circ.*, 226, 12, 1965.

137. Hunter, C.G., Allowable human body concentrations of organochlorine pesticides, *La Medicina del Lavoro*, 59(10), 577, 1968.

138. Hunter, C.G. and Robinson, J., Pharmacodynamics of dieldrin (HEOD). I. Ingestion by human subjects for 18 months, *Arch. Environ. Health*, 15, 614, 1967.

139. Hunter, C.G., Robinson, J., and Richardson, A., Chlorinated insecticide content of human body fat in southern England, *Brit. Med. J.*, 1, 221, 1963.

140. Hunter, C.G., Robinson, J., and Roberts, M., Pharmacodynamics of dieldrin (HEOD). Ingestion by human subjects for 18-24 months and part exposure for 8 months, *Arch. Environ. Health*, 18, 12, 1969.

141. Hurtig, H. and Harris, C.R., Nature and sources of pollution by pesticides, *Conference on Pollution and our Environment Montreal*, 1966, 16.

142. Ide, F.P., Effects of forest spraying with DDT on aquatic insects of salmon streams, *Trans. Amer. Fish Soc.*, 86, 208, 1957.

143. Ide, F.P., Effects of pesticides on stream life, *Develop. Ind. Microbiol.*, 9, 132, 1968.

144. Jacobson, M., The use of attractants and repellants as alternate methods of pest control, in *Research in Pesticides*, Academic Press, New York, 1965, 251.

145. Jefferies, D.J. and Davis, B.N.K., Dynamics of dieldrin in soils, worms and songthrushes, *J. Wildlife Manage.*, 32, 441, 1966.

146. Jefferies, D.J. and Prestt, I., Post-mortem examinations of Peregrine and Lanner falcons with particular reference to organochlorine residues, *Brit. Birds*, 59, 49, 1966.

147. Johnson, D.W., Pesticides and fishes - a review of selected literature, *Trans. Amer. Fish Soc.*, 97, 398, 1968.

148. Jones, J.S. and Hatch, M.B., The significance of inorganic spray residue accumulations in orchard soils, *Soil Sci.*, 44, 37, 1937.

149. Kallman, B.J., Cope, O.B., and Navarre, R.J., Distribution and detoxification of toxaphene in Clayton Lake, New Mexico, *Trans. Amer. Fish Soc.*, 91, 14, 1962.

150. Kearney, P.C., Nash, R.G., and Isensee, A.R., Persistence of pesticide residues in soils, in *Chemical Fallout - Current Research on Persistent Pesticides*, C. C Thomas, Springfield, Ill., 1969, 531.

151. Keith, J.O., Relation of pesticides to water-fowl mortality at Tule Lake Refuge, *U.S. Fish Wildlife Serv., Denver Wildlife Res. Center Ann. Progr. Rept.*, 1964, 14.

152. Keith, J.O., Insecticide contaminations in wetland habitats and their effects on fish-eating birds, *J. Appl. Ecol. (Suppl 3)*, 71, 1966.

153. Keith, J.O. and Hunt, E.G., Levels of insecticide residues in fish and wildlife in California, *Trans. 31st N. Amer. Wildlife Res. Conf.*, 150, 1966.

154. King, S.F., Some effects of DDT on the guppy and the brown trout, *U.S. Fish Wildlife Serv. Spec. Sci. Rept.*, 399, 1962, 22.

155. van Klingeren, B., Koeman, J.H., and van Haaften, J.L., A study on the hare (*Lepus europeus*) in relation to the use of pesticides in a polder in the Netherlands, *J. Appl. Ecol. (Suppl)*, 3, 125, 1966.

156. Knipling, E.F., The sterility method of pest population control, in *Research in Pesticides.*, Academic Press, New York, 1965, 233.

157. Koeman, J.H. and van Genderen, H., Organochlorine residues in birds from the Netherlands, *J. Appl. Ecol. (Suppl)*, 3, 99, 1966.

158. Koeman, J.H., Veen, J., Brouwer, E., Huisman-de-Brouwer, L., and Koolen, J.L., Residues of chlorinated hydrocarbon insecticides in the North Sea environment, *Helgoländer wiss Meersunters*, 17, 375, 1968.

159. Korschgen, L.J. and Murphy, D.A., Pesticide - wildlife relationships: reproduction growth and physiology of deer fed dieldrin - contaminated diets, *Missouri Fed. Aid Proj. 13-R-21 Progr. Rept.*, 1967, 24.

160. Kraybill, H.F., Significance of pesticide residues in foods in relation to total environmental stress, *Can. Med. Ass. J.*, 100, 204, 1969.

161. Lahser, C. and Applegate, H.G., Pesticides at Presidio. III Soil and water, *Texas J. Sci.*, 18(4), 12, 1966.

162. Lamb, D.W., Linder, R.L., and Greichus, Y.A., Dieldrin residues in eggs and fat of penned pheasant hens, *J. Wildlife Manage.*, 31(1), 24, 1967.

163. Langer, E., Pesticides: minute quantities linked with massive fish kills; federal policy still uncertain, *Science*, 144, 35, 1964.

164. Laws, E.R. Jr. and Biros, F.J., Men with intensive occupational exposure to DDT, *Arch. Environ. Health*, 15, 766, 1967.

165. Lichtenstein, E.P., DDT accumulation in mid-western orchard and crop soils treated since 1945, *J. Econ. Entomol.*, 50, 386, 1958.

166. Lichtenstein, E.P., Absorption of some chlorinated hydrocarbon insecticides from soil into various crops, *J. Agr. Food Chem.*, 7(6), 430, 1959.

167. Lichtenstein, E.P., Epoxidation of aldrin and heptachlor in soils as influenced by autoclaving, moisture and soil types, *J. Econ. Entomol.*, 53(2), 192, 1960.

168. Lichtenstein, E.P., Fuhremann, T.W., Scopes, N.E.A., and Skrentny, R.F., Translocation of insecticides from soils into pea plants, *J. Agr. Food Chem.*, 15(5), 864, 1967.

169. Lichtenstein, E.P. and Medler, J.T., Persistence of aldrin and heptachlor residues on alfalfa, *J. Econ. Entomol.*, 51, 222, 1958.

170. Lichtenstein, E.P., Muller, C.H., Myrdal, G.R., and Schulz, K.R., Vertical distribution and persistence of insecticidal residues in soils as influenced by mode of application and a cover crop., *J. Econ. Entomol.*, 55, 215, 1962.

171. Lichtenstein, E.P., Myrdal, G.R., and Schulz, K.R., Absorption of insecticidal residues from contaminated soils into five carrot varieties, *J. Agr. Food Chem.*, 13(2), 126, 1965.

172. Lichtenstein, E.P. and Schulz, K.R., Breakdown of lindand and aldrin in soils, *J. Econ. Entomol.*, 52, 118, 1959.

173. Lichtenstein, E.P. and Schulz, K.R., Effect of soil cultivation, soil surface and water on the persistence of insecticide residues in soils, *J. Econ. Entomol.*, 54, 517, 1961.

174. Lichtenstein, E.P. and Schulz, K.R., Residues of aldrin and heptachlor in soils and their translocation into various crops, *J. Agr. Food Chem.*, 13(1), 57, 1965.

175. Lichtenstein, E.P., Schulz, K.R., Skrentny, R.F., and Tsukano, Y., Toxicity and fate of insecticide residues in water, *Arch. Environ. Health*, 12, 199, 1966.

176. Lichtenstein, E.P., Fuhremann, T.W., and Schulz, K.R., Use of carbon to reduce the uptake of insecticidal soil residues by crop plants, *J. Agr. Food Chem.*, 16(2), 348, 1968.

177. Loosanoff, V.L., Pesticides in sea water and the possibilities of their use in mariculture, in *Research in Pesticides*, Academic Press, New York, 1965, 135.

178. Lotse, E.G., Graetz, D.A., Chesters, G., Lee, G.B., and Newland, L.W., Lindane adsorption by lake sediments, *Env. Sci. Technol.*, 353, 1968.

179. Lowden, G.F., Saunders, C.L., and Edwards, R.W., Organochlorine insecticides in water (Pt. II), *Proc. Soc. Water Treat. Exam.*, (in press)

180. Lyman, L.D., Tompkins, W.A., and McCann, J.A., Massachusetts pesticide monitoring study, *Pest. Mon. J.*, 2(3), 109, 1968.

181. Mack, G.L., Corcoran, S.M., Gibbs, S.D., Gutenmann, W.H., Reckahn, J.A., and Lisk, D.J., The DDT content of some fishes and surface waters of New York State, *N.Y. Fish Game J.*, 11(2), 148, 1964.

182. MacRae, I.C., Raghu, K., and Castro, T.F., Persistence and biodegradation of four common isomers of benzene hexachloride in submerged soils, *J. Agr. Food Chem.*, 15, 911, 1967.

183. Maier-Bode, H., Insecticidal residues resulting from treatment of cherries with dichlorodiphenyltrichloroethane and methoxychlor to control the cherry fruit fly, *Z Pflanzenkr Pflanzenpathol Pflanzenschutz*, 68, 261, 1961.

184. Marth, E.H., Residues and some effects of chlorinated hydrocarbon insecticides in biological material, *Residue Reviews*, 9, 1, 1965.

185. Matsumura, F. and Boush, G.M., Degradation of insecticides by a soil fungus, *J. Econ. Entomol.*, 61, 610, 1968.

186. Mellanby, K., *Pesticides and pollution*, New Naturalist, No. 50, Collins, London, 1967, 221.

187. Middleton, F.M. and Lichtenberg, J.J., Measurement of organic contaminants in the nation's rivers, *Ind. Eng. Chem. Eng. Anal. Ed.*, 52(6), 99, 1960.

188. Miller, N.W. and Berg, G.G., *Chemical Fallout—Current Research on Persistent Pesticides.*, C.C Thomas, Springfield, Ill., 1969, 531.

189. Miller, L.A., Miles, J.R.W., and Sans, W.W., DDT and DDD residues on tomatoes processed into juice, *Can. J. Plant Sci.*, 37, 288, 1957.

190. Ministry of Agriculture, Fisheries & Food, *Review of the persistent organochlorine pesticides*, HMSO, London, 1964, 68.

190a. Ministry of Agriculture, Fisheries and Food, *Further Review of Certain Organochlorine Pesticides Used in Great Britain*, HMSO, London, 1969, 148.

191. Moore, N.W., Pesticides and birds - a review of the situation in Great Britain in 1965, *Bird Study*, 12, 222, 1965.

192. Moore, N.W., *Pesticides in the Environment and their Effects on Wildlife*, Blackwell, Oxford, 1966, 311.

193. Moore, N.W., Reduction of pesticide hazards to wildlife - an appraisal of experience gained in Great Britain 1960-7, *Proc. Brit. Ins. Fung. Conf.*, 2, 493, 1967.

194. Moore, N.W., A synopsis of the pesticide problem, *Adv. Ecol. Res.*, 4, 75, 1967.

195. Moore, N.W. and Tatton, J. O'G., Organochlorine insecticide residues in the eggs of sea birds, *Nature*, London, 207, 42, 1965.

196. Moriarty, F., The sublethal effects of synthetic insecticides on insects, *Biol. Rev.*, 44, 321, 1969.

197. Morrison, H.E., Crowell, H.H., Crumb, S.E., and Lauderdale, R.W., The effects of certain new soil insecticides on plants, *J. Econ. Entomol.*, 41, 374, 1948.

198. Mount, D.I. and Putnicki, G.J., Summary report of the 1963 Mississippi fish kill, *Trans. 31st N. Amer. Wildlife Nat. Res. Conf.*, 177, 1966.

199. Moye, W.C. and Luckmann, W.H., Fluctuations in populations of certain aquatic insects following application of aldrin granules to Sugar Creek, Iroquois County Illinois, *J. Econ. Entomol.*, 57(3), 318, 1964.

200. Murphy, R.T., Fahey, J.E., and Miles, E.J., DDT residues in southern Indiana orchard soils in 1963, *Proc. N. Centr. Bran. ESA*, 19, 144, 1964.

201. Nash, R.G. and Woolson, E.A., Persistence of chlorinated hydrocarbon insecticides in soils, *Science (NY)*, 157, 924, 1967.

202. Newland, L.W., Chesters, G., and Lee, G.B., Degradation of γ-BHC in simulated lake impounds as affected by aeration, *J. Water Poll. Control Fed.*, (2), 174, 1969.

203. Newsom, L.D., Consequences of insecticide use on non-target organisms, *Ann. Rev. Entomol.*, 12, 257, 1967.

204. Nicholson, H.P., Pesticide pollution control, *Science (NY)*, 158(3803), 871, 1967.

205. Nicholson, H.P., Occurrence and significance of pesticide residues in water, *Proc. Wash. Acad. Sci.*, 59(4-5), 77, 1969.

206. Nicholson, H.P., Grzenda, A.R., and Teasley, J.I., Water pollution by insecticides. A six and one-half year study of a water shed, *Proc. Symp. Agr. Waste Water*, 132, 1966.

207. Perkow, W., *Die Insektizide*, Hüther Verlag, Heidelberg, 1956, 360.

208. •Prestt, J., Investigations into possible effects of organochlorine insecticides on wild predatory birds, *Proc. 4th Brit. Ins. Fung. Conf.*, 26, 1967.

209. Pillmore, R. and Finley, R.B. Jr., Residues in game animals resulting from forest and range insecticide applications, *Trans. 28th N. Amer. Wildlife Conf.*, 28, 409, 1963.

210. Ratcliffe, D.A., Organochlorine residues in raptor and corvid eggs from northern Britain, *Brit. Birds*, 58(3), 65, 1965.

211. Reichel, W.L. and Addy, C.E., A survey of chlorinated pesticide residues in black duck eggs, *Bull. Environ. Contam. Toxicol.*, 3(3), 174, 1968.

212. Reichel, W.L., Lamont, T.G., Cromartie, E., and Locke, L.N., Residues in two bald eagles suspected of pesticide poisoning, *Bull. Environ. Contam. Toxicol.*, 4(1), 24, 1969.

213. Reymonds, T.D., Pollution effects of agricultural insecticide and synthetic detergents, *Water Sewage Works*, 109, 352, 1962.

214. Riseborough, R.W., Chlorinated hydrocarbons in marine ecosystems, in *Chemical Fallout - Research on Persistent Pesticides*, C.C Thomas, Springfield, Ill., 1969, 531.

215. Riseborough, R.W., Huggett, R.J., Griffin, J.J., and Goldberg, E.D., Pesticides: Transatlantic movements in the north east Trades, *Science (NY)*, 159, 1233, 1968.

216. Robbins, C.S., Springer, P.F., and Webster, C.G., Effects of five-year DDT application on breeding bird population, *J. Wildlife Manage.*, 15, 213, 1951.

217. Robeck, G.G., Effectiveness of water treatment processes in pesticide removal, *J. Amer. Water Works Ass.*, 57, 181, 1965.

218. Robinson, J., Organochlorine insecticides and birds, *Chem. Brit.*, 4(4), 158, 1968.

219. Robinson, J., Residues of organo-chlorine insecticides in dead birds in the U.K., *Chem.*, London, 47, 1974, 1967.

220. Robinson, J., Organochlorine insecticides and bird populations in Britain, in *Chemical Fallout - Current Research on Persistent Pesticides*, C. C Thomas, Springfield, Ill., 1969, 113.

221. Robinson, J., The burden of chlorinated hydrocarbon pesticides in man, *Can. Med. Ass. J.*, 100, 180, 1969.

222. Robinson, J. and McGill, A.E.J., Organochlorine insecticide residues in complete prepared meals in Great Britain during 1965, *Nature*, London, 212, 1037, 1966.

223. Robinson, J., Richardson, A., Hunter, C.G., Crabtree, A.N., and Rees, H.J., Organochlorine insecticide content in human adipose tissue in southeastern England, *Brit. J. Ind. Med.*, 22, 220, 1965.

224. Robinson, J., Richardson, A., Crabtree, A.N., Coulson, J.C., and Potts, G.R., Organochlorine residues in marine organisms, *Nature*, London, 214(5095), 1307, 1967.

225. Robinson, J. and Roberts, M., Accumulation, distribution and elimination of organochlorine insecticides by vertebrates, *Symposium on Physicochemical and Biophysical Factors Affecting the Activity of Pesticides*, 106, 1967.

226. Robinson, J. and Roberts, M., The estimation of the exposure of the general population to dieldrin, *J. Food Cosmet. Toxicol.*, (in press)

227. Robinson, J., Roberts, M., Baldwin, M., and Walker, A.I.T., The pharmacokinetics of HEOD (dieldrin) in the rat, *J. Food Cosmet. Toxicol.*, (in press)

228. Rosen, A.A. and Middleton, F.M., Chlorinated insecticides in surface waters, *Anal. Chem.*, 31, 1729, 1959.

229. Rudd, R.L., *Pesticides and the Living Landscape*, Faber & Faber, London, 1964, 320.

230. Saha, J.G., Craig, C.H., and Janzen, W.K., Organochlorine residues in agricultural soil and legume crops in N.E. Saskatchewan, *J. Agr. Food Chem.*, 16(4), 617, 1968.

231. Saha, J.G. and McDonald, M., Insecticide residues in wheat grown in soil treated with aldrin and endrin, *J. Agr. Food Chem.*, 15, 205, 1967.

232. Saha, J.G. and Stewart, W.W.A., Heptachlor, heptachlor-epoxide and gamma chlordane residues in soil and rutabaga after soil and surface treatment with heptachlor, *Can. J. Plant Sci.*, 47, 79, 1967.

233. Seal, W.L., Dowsey, L.H., and Cavin, G.E., Pesticides in soil. Monitoring for chlorinated hydrocarbon pesticides in soil and root crops in the Eastern States in 1965, *Pest. Mon. J.*, 1(3), 22, 1967.

234. Simmons, S.W., Tests of the effectiveness of DDT in anopheline control, *Publ. Health Rept.*, 60, 917, 1945.

235. Sladen, W.J.L., Menzie, C.M., and Reichel, W.L., DDT residues in Adelie penguins and a crabeater seal from Antarctica. Ecological implications, *Nature*, London, 210(5037), 670, 1966.

236. Sparr, B.I., Appleby, W.G., DeVries, D.M., Osmun, J.V., McBride, J.M., and Foster, G.L., Insecticide residues in waterways from agricultural use, *Adv. Chem. Ser.*, 60, 146, 1966.

237. Stickel, L.F., Organochlorine pesticides in the environment, *U.S.D.I. Bur. Sport Fish Wildlife Rept.*, 119, 1, 1968.

238. Stickel, W.H., Dodge, W.E., Sheldon, W.G., Dewitt, J.B., and Stickel, L.F., Body condition and response to pesticides in woodcocks, *J. Wildlife Manage.*, 29(1), 147, 1965.

239. Stickel, W.H., Hayne, D.W., and Stickel, L.F., Effects of heptachlor contaminated earthworms on woodcocks, *J. Wildlife Manage.*, 29(1), 132, 1965.

240. Stickel, L. and Stickel, W.H., Distribution of DDT residues in tissues of birds in relation to mortality, body condition and time, *Ind. Med. Surg.*, 38(3), 44, 1969.

241. Stickel, W.H., Stickel, L.F., and Spann, J.W., Tissue residues of dieldrin in relation to mortality in birds and mammals, in *Chemical Fallout; Current Research on Persistent Pesticides*, C.C Thomas, Springfield, Ill., 1968.

242. Stickel, L.F., Chura, N.J., Stewart, P.A., Menzie, C.M., Prouty, R.M., and Reichel, W.L., Bald eagle pesticide relations, *Trans. 31st N. Amer. Wildlife Nat. Res. Conf.*, 190, 1966.

243. Street, J.C., Ecological systems: domestic animals, in *Research in Pesticides*, Academic Press, New York, 1965, 151.

244. Street, J.C., Methods of removal of pesticide residues, *Can. Med. Ass. J.*, 100, 154, 1969.

245. Strickland, A.H., Amounts of organochlorine insecticides used annually on agricultural and some horticultural crops in England and Wales, *Ann. Appl. Biol.*, 55, 319, 1965.

246. Strickland, A.H., Some estimates of insecticide and fungicide usage in agriculture and horticulture in England and Wales, 1960-4, *J. Appl. Ecol. (Suppl)*, 3, 3, 1966.

247. Stringer, A. and Pickard, J.A., The DDT content of soil and earthworms in an apple orchard at Long Ashton, *Rept. Agr. Hort. Res. Sta. Univ. Bristol for 1962*, 127, 1963.

248. Tabor, E.C., Contamination of urban air through the use of insecticides, *Trans. N.Y. Acad. Sci. Ser.*, 2, 28(5), 569, 1966.

249. Tarrant, K.R. and Tatton, J. O'G., Organochlorine pesticides in rainwater in the British Isles, *Nature*, London, 219(5155), 725, 1968.

250. Terriere, L.C., Kiigemagi, V., Gerlach, A.R., and Boronicka, R.L., The persistence of toxaphene in lake water and its uptake by aquatic plants and animals, *J. Agr. Food Chem.*, 14, 66, 1965.

251. Tew, R.P. and Sillibourne, J.M., Pesticide residues on fruit. IV Endrin residues on black currants, *J. Sci. Food Agr.*, 12, 661, 1961.

252. Thompson, A.R., Edwards, C.A., Edwards, M.J., and Beynon, K.I., The movement of dieldrin through soil. II. In sloping troughs and soil columns, *J. Sci. Food Agr.*, (in press)

253. Trautmann, W.L., Organochlorine insecticide composition of randomly selected soils from nine States 1967, *Pest. Mon. J.*, 2(2), 93, 1968.

254. Tsukano, Y. and Suzuki, T., Absorption and translocation of BHC by rice plants, *Botyo-Kagaku,* 27, 12, 1962.

255. Tu, C.M., Miles, J.R.W., and Harris, C.R., Soil microbial degradation of aldrin, *Life Sci.*, 7, 311, 1968.

256. Turner, N., DDT in Connecticut wildlife, *Conn. Agr. Exp. Sta. Bull.*, 672, 1, 1965.

257. Turtle, E.E., Taylor, A., Wright, E.N., Thearle, R.J.P., Egan, H., Evans, W.H., and Soutar, N.M., The effects on birds of certain chlorinated insecticides used as seed dressings, *J. Sci. Food Agr.*, 8, 567, 1963.

258. U.S.D.A., Report of Committee on persistent pesticides, 1969, 33.

259. U.S.D.A., Monitoring agricultural pesticide residues, *U.S.D.A. Publ., ARS 81-13*, Washington, 1966, 53.

260. U.S.D.A., Monitoring for chlorinated hydrocarbon insecticide residues in soybeans - 1966, *Pest. Mon. J.*, 2(1), 58, 1968.

261. U.S. Department of Health, Education & Welfare, *Pesticides in Soil and Water*, 1964, 90.

262. U.S.D.I., Pesticide—Wildlife Studies 1962—A Review of Fish and Wildlife Service Investigations During the Calendar Year., *U.S.D.I. Fish Wildlife Serv. Circ.*, 167, 1963, 109.

263. U.S.D.I., Pesticide-Wildlife Studies 1963. A Review of Fish & Wildlife Service Investigations During the Calendar Year., *U.S.D.I. Fish Wildlife Serv. Circ.*, 199, 1964, 129.

264. U.S.D.I., The effects of pesticides on fish and wildlife, 1964, *U.S.D.I. Fish Wildlife Serv. Circ.*, 226, 1965, 77.

265. U.S.D.I., Wildlife Research Problems, Programs, Progress, 1965, *U.S.D.I. Fish Wildlife Serv. Bur.-Sport Fish Wildlife*, 1966, 102.

266. U.S.D.I., Wildlife Research Problems, Programs, Progress, 1966, *U.S.D.I. Fish Wildlife Serv. Bur.-Sport Fish Wildlife,* 1967, 116.

267. del Vecchio, V. and Leoni, V., La ricerva ed il dosaggio degli insetticidi clorurati in materiale biologico, *Nuovi Ann. Ig. Microbiol.*, 28, 107, 1967.

268. Vlieger, M de., Robinson, J., Baldwin, M.K., Crabtree, A.N., and Dijk, M.C. van., The organochlorine insecticide content of human tissues, *Arch. Environ. Health*, 17, 759, 1968.

269. Waites, R.E. and van Middelm, C.H., Residue studies of DDT and malathion on turnip tops, collards, snap beans and lettuce, *J. Econ. Entomol.*, 51, 306, 1958.

270. Walker, C.H., Insecticide and herbicide residues in soil, *Monks Wood Exp. Sta. Rept. for 1960-5, Nature Conservancy, London*, 55, 1966.

271. Walker, C.H., Some chemical aspects of residue studies with DDT, *J. Appl. Ecol. (Suppl)*, 3, 213, 1966.

272. Walker, C.H., Hamilton, G.A., and Harrison, R.B., Organochlorine insecticide residues in wild birds in Britain, *J. Sci. Food Agr.*, 18, 123, 1967.

273. Walker, K.C., George, D.A., and Maitlen, J.C., Residues of DDT in fatty tissues of big game animals in the States of Idaho and Washington in 1962, *U.S.D.A. ARS 33 - 105*, 21, 1965.

274. Ware, G.W., Estesen, B.J., and Cahill, W.P., An ecological study of DDT residues in Arizona soils and alfalfa, *Pest. Mon. J.*, 2(3), 129, 1968.

275. Warnick, S.L., Gaufin, R.F., and Gaufin, A.R., Concentrations and effects of pesticides in aquatic environments, *J. Amer. Water Works Ass.*, 58, 601, 1966.

276. Wasserman, M., Wasserman, D., Zellermayer, L., and Gon, M., Storage of DDT in the people of Israel, *Pest. Mon. J.*, 1(2), 15, 1967.

277. Weaver, L., Gunnerson, C.G., Breidenbach, A.W., and Lichtenberg, J.J., Chlorinated hydrocarbon pesticides in major U.S. river basins, *Publ. Health Rept., Washington*, 80, 481, 1965.

278. Webb, R.E. and Horsfall, F. Jr., Endrin resistance in the pine mouse. *Science (NY)*, 156(3783), 1762, 1967.

279. Weibel, S.R., Weidner, R.B., Cohen, J.M., and Christianson, A.G., Pesticides and other contaminants in rainfall and runoff, *J. Amer. Water Works Ass.*, 58, 1075, 1966.

280. Weidhass, D.E., Schmidt, C.H., and Bowman, M.C., Effects of heterogeneous distribution and codistillation on the results of tests with DDT against mosquito larvae, *J. Econ. Entomol.*, 53(1), 121, 1960.

281. Wene, G.P., Toxaphene residues on certain vegetables at various time intervals after application, *J. Rio Grande Valley Hort. Soc.*, 12, 106, 1958.

282. Wheatley, G.A. and Hardman, J.A., Indications of the presence of organochlorine insecticides in rainwater in central England, *Nature*, London, 207, 486, 1965.

283. Wheatley, G.A. and Hardman, J.A., Organochlorine insecticide residues in earthworms from arable soils, *J. Sci. Food Agr.*, 19, 219, 1968.

284. Wheatley, G.A., Hardman, J.A., and Strickland, A.H., Residues of chlorinated hydrocarbon insecticides in some farm soils in England, *Plant Pathol.*, 11, 81, 1962.

285. Whitten, J.L., *That We May Live*, van Nostrand, New York, 1966, 251.

286. Wiese, I.H. and Basson, N.C.J., The degradation of some persistent chlorinated hydrocarbon insecticides applied to different soil types, *S. Afr. J. Agr. Sci.*, 9, 945, 1966.

287. Williams, S.P., Mills, A., and McDowell, R.E., Residues in milk of cows fed rations containing low concentrations of five chlorinated hydrocarbon pesticides, *J. Ass. Off. Anal. Chem.*, 47(6), 1124, 1964.

288. Wingo, C.W., Persistence and degradation of dieldrin and heptachlor in soil and effects on plants, *Univ. Missouri Agr. Exp. Sta. Res. Bull.*, 914, 1966, 27.

289. Wit, S.L., Enige Aspecten van de Toxicologie en Chemische analyse van bestrijdingsmiddelen-residus, *Voeding*, 25, 609, 1964.

290. Wolfe, H.R., Elliott, J.W., and Durham, W.F., The trend of DDT and parathion residues on apples grown in central Washington, *J. Econ. Entomol.*, 52, 1053, 1959.

291. Woodwell, G.M. and Martin, F.T., Persistence of DDT in soils of heavily sprayed forest stands, *Science (NY)*, 145, 481, 1964.

292. World Health Organization, Pesticide residues in food, *WHO & FAO Tech. Rept.*, 370, 19, 1967.

293. World Health Organization, Pesticide residues, *WHO & FAO Tech. Rept.*, 391, 43, 1968.

294. Yaron, B., Swoboda, A.R., and Thomas, G.W., Aldrin adsorption by soils and clays, *J. Agr. Food Chem.*, 15(4), 671, 1967.

295. Young, L.A. and Nicholson, H.P., Stream pollution resulting from the use of organic insecticides, *Progr. Fish Cult.*, 13, 193, 1951.

296. Yule, W.N., DDT residues in forest soils, *In litt*, 1969.

297. Zavon, M.R., Hine, C.H., and Parker, K.D., Chlorinated hydrocarbon insecticides in human body fat in the United States, *J.A.M.A.*, 193, 837, 1965.

298. Anon., Restoring the quality of our environment, Report of the Environmental Pollution Panel, President's Science Advisory Committee (The White House), Nov. 1965.

299. Borg, K., Wanntorp, K.E., and Hanko, E., Mercury poisoning in Swedish Wildlife. *J. Appl Ecol.*, (Suppl), 3, 171, 1966.

300. Brown, A.W.A., Insecticide resistance-genetic implications and applications, *World Rev. Pest Contr.*, 6(4), 104, 1967.

301. Burnside, O.C., Fenster, C.R., and Wicks, G.A., Dissipation and leaching of monuron, simazine and atrazine in Nebraska soils, *Weeds,* 11, 209, 1963.

302. Busvine, J.R., Insecticide resistance and the future of pest control, *PANS* 14(3), 311, 1968.

303. Dalton, R.L., Evans, A.W., and Rhodes, R.C., Disappearance of diuron from cotton field soils, *Weeds,* 14, 31, 1966.

304. Department of Education and Science, Further review of certain persistent organochlorine pesticides used in Great Britain, *Report by the Advisory Committee on Pesticides and other Toxic Chemicals,* HMSO London, 1969, 148.

305. Dowler, C.C., Sand, P.F., and Robinson, E.L., The effect of soil type on pre-planting soil-incorporated herbicides for witchweed control, *Weeds,* 11, 276, 1963.

306. Faust, S.D., Pollution of the water environment by organic pesticides, *Clin. Pharmacol. Ther.,* 5(6), 677, 1964.

307. Faust, S.D., and Aly, O.M., Water pollution by organic pesticides. *J. Amer. Water Works Ass.,* 56, 267, 1964.

308. Fletcher, W.W., The effect of herbicides on soil microorganisms, in *Herbicides and the Soil,* Woodford, E.K., and Sagon, G.R., Eds., Blackwell, Oxford, 1960, 41.

309. Fletcher, W. W., Effect of organic herbicides on soil microorganisms, *Pest Tech.,* 1961, 272.

310. Grolleau, G. and Biaddi, F., Note on the effects of Thiram on the laying and rearing of the red-legged partridge (*Alectoris rufa*), *J. Appl. Ecol.,* 3, 249, 1966.

311. Holden, A., The possible effects on fish of chemicals used in agriculture, *J. Proc. Inst. Sewage Purif.,* 1964, 361.

312. Jones, J.R.E., *Fish and River Pollution,* Butterworth s, London, 1964.

313. Phillips, W.M., Residual herbicidal activity of some chloro-substituted benzoic acids in soil, *Weeds,* 7, 284, 1959.

314. Raw, F. and Lofty, J.R., Earthworm populations in orchards, *Rep. Rothamsted Exp. Sta. for 1959,* 134, 1960.

315. Sheets, T. J., The extent and seriousness of pesticide buildup in soils, in *Agriculture and the Quality of our Environment,* A.A.A.S. Publ. 85, 1967, 20.

316. Schweizer, E.E., Toxicity of DSMA soil residues to cotton and rotational crops, *Weeds,* 15, 72, 1967.

317. U.S. Department of Agriculture, *Report of Committee on Persistent Pesticides,* Division of Biology and Agriculture, National Research Council to U.S.D.A., Washington, D.C., May 1969, 34.

318. U.S. Department of Health, Education, and Welfare, *Report of the Secretary's Commission on Pesticides and their Relationship to Environmental Health,* Pts. I & II. U.S. Govt. Printing Office, Washington, D.C., 1969, 677.

319. van Valin, C.C., Persistence of 2, 6-dichlorobenzonitrile in aquatic environments, in *Organic Pesticides in the Environment, Advances in Chemistry Series,* 66, 217, 1966.

320. Vos, J.G., Breeman, H.A., and Benschop, H., The occurrence of the fungicide hexachlorobenzene in wild birds and its toxicological importance. A preliminary communication, *Med. Rijk. Landbouw Wet. Gent.,* 33(3), 1263, 1968.

321. Williams, J.H., Herbicides—their fate and persistence in soils, *N.A.A.S. Quart. Rev.,* 87, 119, 1970.

322. Woodwell, G.M., Toxic substances and ecological cycles, *Sci. Amer.,* 216, 1967.